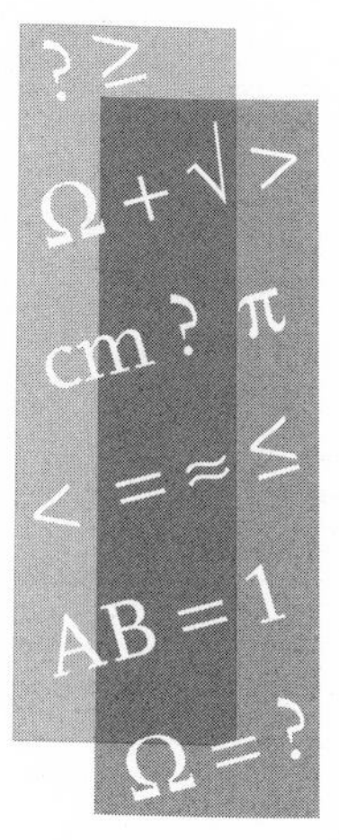

Brainteaser Physics

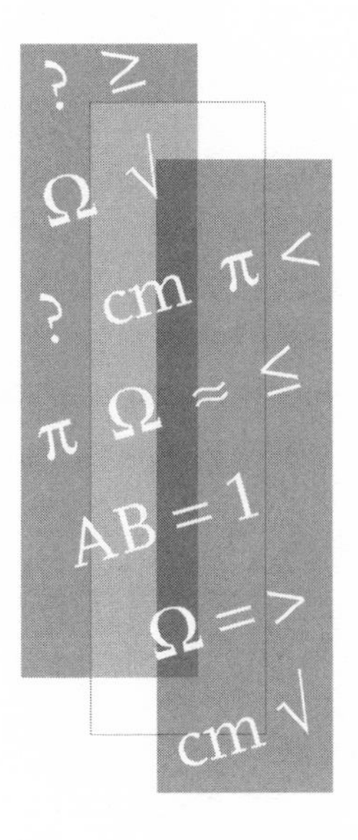

Brainteaser Physics

Challenging Physics Puzzlers

Göran Grimvall

The Johns Hopkins University Press
Baltimore

Printed in the United States of America on acid-free paper
9 8 7 6 5 4 3 2 1

The Johns Hopkins University Press
2715 North Charles Street
Baltimore, Maryland 21218-4363
www.press.jhu.edu

Library of Congress Cataloging-in-Publication Data

Grimvall, Göran.
Brainteaser physics: challenging physics puzzlers / Göran Grimvall.
p. cm.
Includes bibliographical references and index.
ISBN-13: 978-0-8018-8511-2 (acid-free paper)
ISBN-10: 0-8018-8511-6 (acid-free paper)
ISBN-13: 978-0-8018-8512-9 (pbk. : acid-free paper)
ISBN-10: 0-8018-8512-4 (pbk. : acid-free paper)
1. Physics—Miscellanea. I. Title.
QC75.G75 2007
530—dc22 2006025675

A catalog record for this book is available from the British Library.

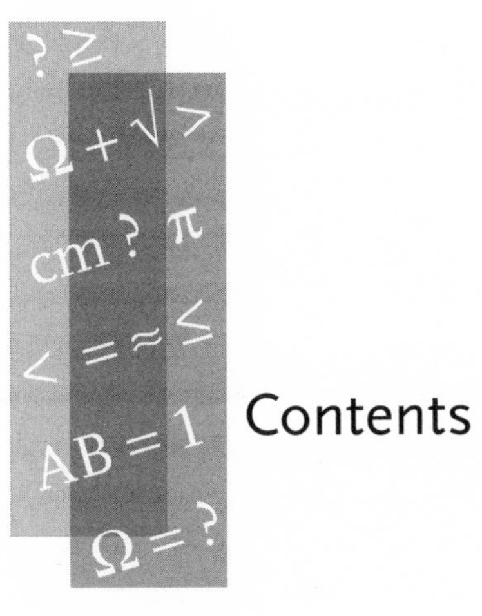

Contents

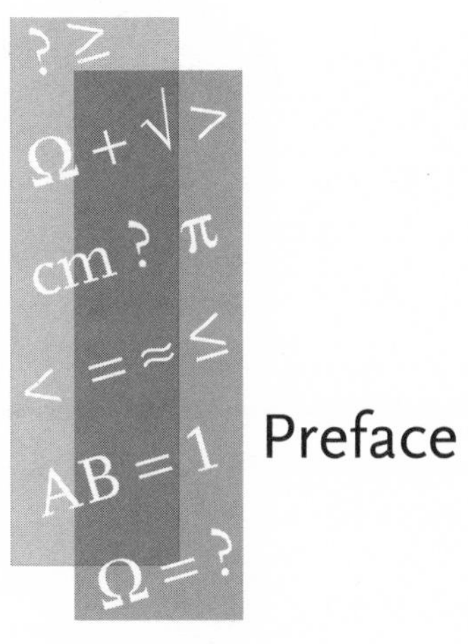

Preface

Solving problems and coping with challenges are human traits. Many of us do these things just for fun. Crosswords and chess problems are found virtually everywhere. Recreational mathematics problems is a genre with a vast literature. Much less common are recreational physics problems—the theme of this book. Here I present 57 problems. Some of them are well known in the popular scientific literature. Others are classics that have been treated in the pedagogical physics literature. References to such works are given at the end of the book. Most of the problems have appeared in shorter versions in my weekly column, which has been running for more than 27 years in a Swedish journal for engineers. I am presenting them now for the first time for an international audience. I don't claim originality for all those problems, but many of them are given a new twist.

The problems in this book have two sides. One provides a challenge—just for fun or recreation. The other is more serious—it shows how physicists think and thus offers training that could also be of professional use. So, the level of discussion in the solutions varies. It can be elementary in the simplest problems, which can be solved without much knowledge of physics. In the more difficult problems, though, a certain background in mathematics is assumed.

The book concludes with some thoughts about how easy it is to make mistakes. Most likely there are several imperfections or outright errors in this book. All comments from readers are welcome. They can be sent to me at the AlbaNova University Center, Royal Institute of Technology, Stockholm, Sweden.

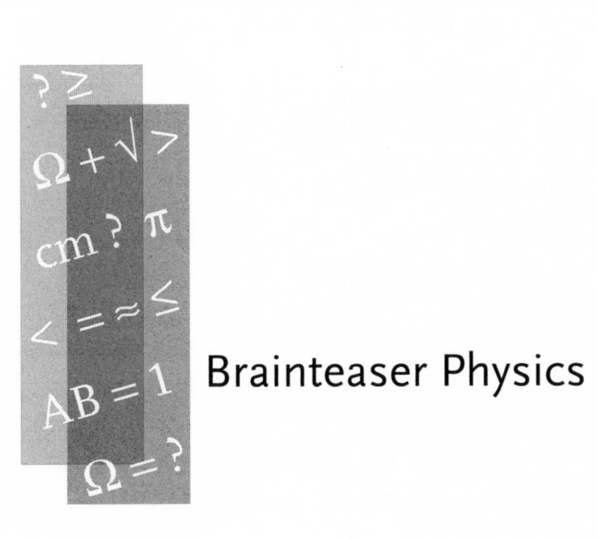

Brainteaser Physics

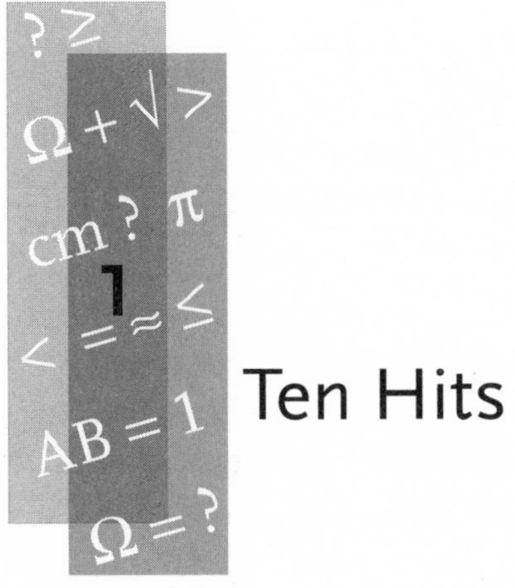

Ten Hits

For well over a century the popular scientific literature has included braintwisters with solutions that are applications of simple physical ideas. Many of these problems have become classics and have been published over and over again. Some classic problems that require more of mathematics are found in introductory physics textbooks. This chapter contains ten of these "hits," roughly in order of increasing difficulty.

PROBLEMS

1.1 Dinghy in Pool

A dinghy, with a passenger and an anchor, floats in a small pool. The passenger throws the anchor into the water. Will the water level in the pool increase, decrease, or stay the same?

1.2 Ice in Water

Ice cubes float in a glass, filled to the brim with water. Will the water spill over when the ice melts?

1.3 Accident in Aqueduct

An aqueduct for boats is like a bridge and canal combined. It rests on supports at its ends. When a barge carrying a tractor is in the middle of the aqueduct, the tractor unfortunately rolls into the water and sinks, while the barge still floats. The mass of the tractor is one ton and the mass of the barge alone is five tons. Is the total force on the supports at the ends of the aqueduct changed? If so, by how much?

1.4 Floating Candle

A candle, originally 8 cm long, decreases in length by 2 cm/h when it is burning in an ordinary candlestick. Let it now float in water instead (see fig. 1.1). (A small iron nail at the bottom of the candle lowers the center of gravity so that the candle floats with a vertical orientation.) At the start, the top of the candle is 1 cm above the water level. Is the candle extinguished in less than one hour by the water in the glass?

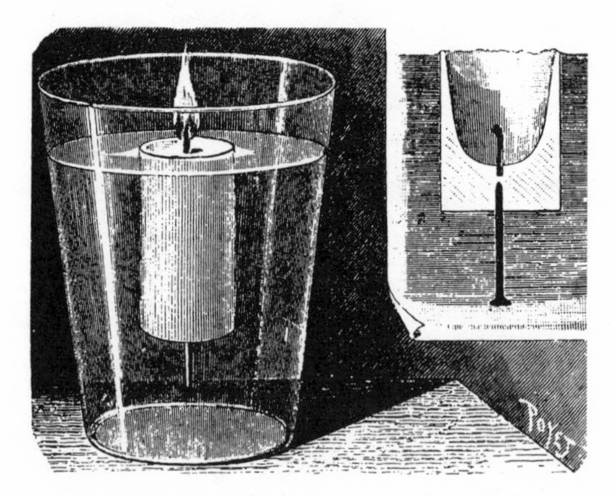

Fig. 1.1. A candle floats in a glass of water. When is it extinguished?

1.5 Running in the Rain

It is raining and you must cross the large parking lot to your car. Because you don't have proper clothes for the rain, and no umbrella, you wonder if it's best to walk or to run to your car. If you walk, you will be exposed to the rain much longer, but running "into the rain" may not be better. What will you do?

1.6 Reaching Out

You have a set of hardcover books that are all the same size. Put one book at the edge of the table, with a certain overhang. Add more books on top of one another (fig. 1.2). What is the minimum number of books needed to make the overhang of the top book clear the edge of the table?

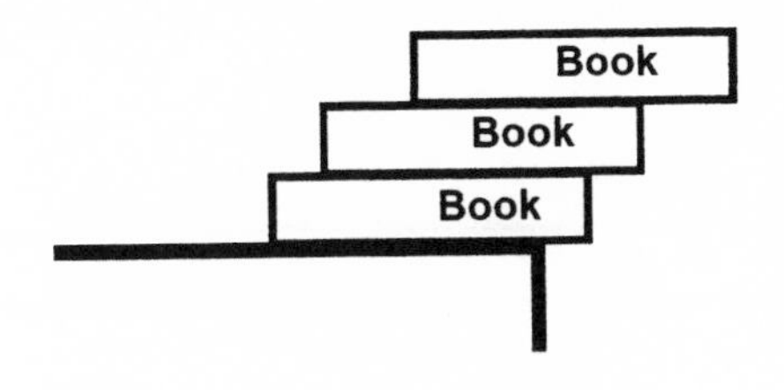

Fig. 1.2. How many books, stacked in a leaning pile, are needed for the top book to clear the edge of the table?

1.7 Resistor Cube

Twelve resistors, all with the resistance 12 Ω, are connected to form the edges of a cube (fig. 1.3). What is the largest resistance between any two corners in the cube?

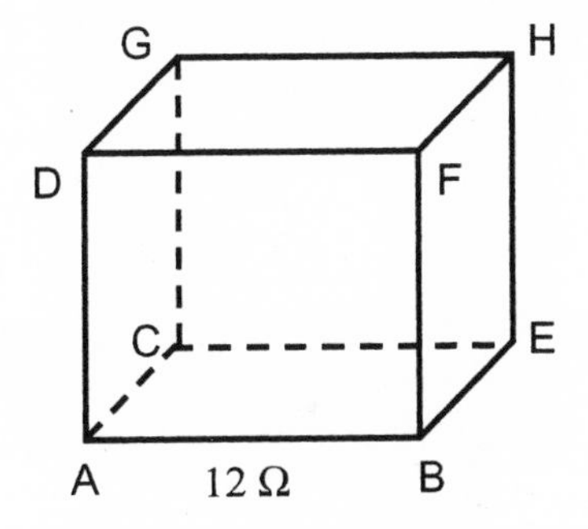

Fig. 1.3. What is the largest resistance between any corners in this cube? Each link has the resistance 12 Ω.

1.8 One, Two, Three, Infinity

A resistor network has the form of a long ladder (fig. 1.4). Each step, and each section of the legs between the steps, is represented by a resistor labeled 1 kΩ. What is the total resistance, as measured between A and B in the figure, if the ladder has *seven* steps?

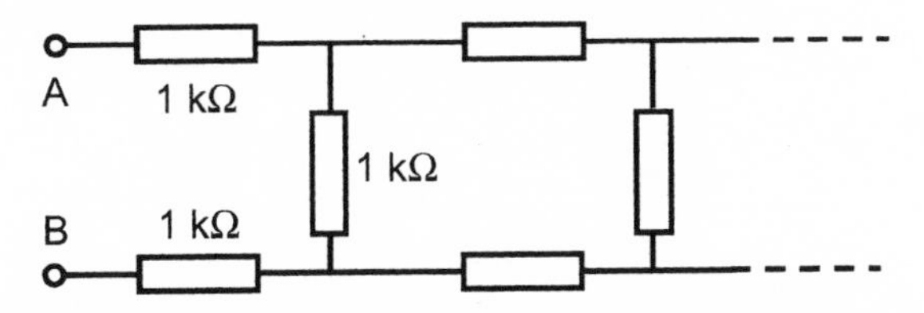

Fig. 1.4. Resistor "ladder" with two steps. What is the resistance between A and B in a ladder with seven steps?

1.9 Lost Energy

You have two capacitors with equal capacitance C. One is given the charge Q and the other is uncharged. They are connected as in figure 1.5. When the two switches are closed, the charge Q will be equally divided between the capacitors.

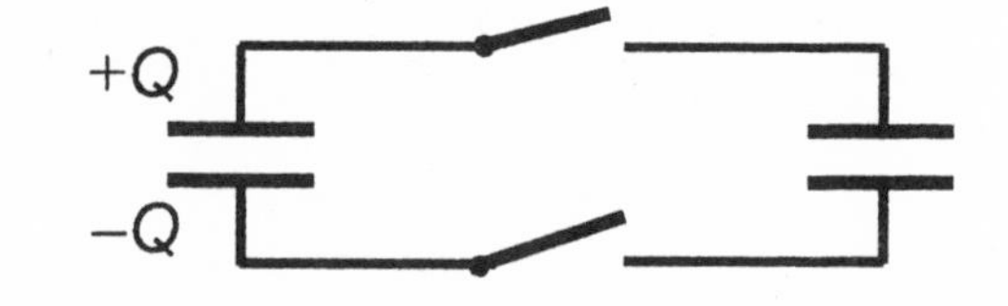

Fig. 1.5. The switches are closed so that a charged capacitor is connected with an uncharged capacitor.

The well-known formula for the energy of a capacitor is

$$E = \frac{Q^2}{2C}$$

This is also the total energy of the two capacitors before the connection is made. When the switches are closed, the charge Q becomes equally divided between the capacitors. The total energy is

$$\frac{(Q/2)^2}{2C} + \frac{(Q/2)^2}{2C} = \frac{Q^2}{4C}$$

Half of the energy that was originally stored in the charged capacitor seems to have disappeared. Where has it gone?

1.10 Simple Timetable

Ignore all practical objections and imagine a transportation system in which every capital on the Earth is connected with every other capital through straight tunnels. All trains depart on the hour, that is, at 8:00, 9:00, 10:00, and so on. When do they arrive at their destinations, if the trains slide through the tunnels under the influence of gravity and without any resistance? The Earth is considered to be a homogeneous, nonrotating sphere.

SOLUTIONS

1.1 Dinghy in Pool

The level decreases, but the effect is *very* small.

This problem is ideal for a consideration of really extreme limits. Let the mass of the dinghy plus the passenger be negligibly small. Furthermore, let the anchor be as small as a pea but made of a material with an extremely high density. In fact, the mass of the anchor is so large that the dinghy barely floats. Water with a volume equal to that of the dinghy is displaced. That raises the water level in the pool.

Next we throw the heavy but miniscule anchor overboard. Because it is so small, we can ignore its effect on the water level, when it lies on the pool bottom. The dinghy with its passenger, assumed to have negligible mass, now floats high without displacing any water. Therefore less water is displaced when the anchor is thrown overboard. As a consequence the water level in the pool is lowered.

Is it realistic? The problem arguably is the most popular physics braintwister. But is it realistic? How much would the water level increase? In the calculation below we see that the lowering of the water level is very small—typically about 1 mm (1/24 in) in a garden pool. This is about the same as the decrease in the water level due to the thermal contraction of the water, if the temperature drops from 25 °C to 22 °C (about 77 °F to 72 °F), and the depth of the pool is 1.5 m (5 ft).

We can make the following estimation in SI units. Suppose that the anchor's mass is 15 kg. The buoyancy force on a floating object is equal to the weight of the fluid that is displaced by the object (*Archimedes' principle*). Before the anchor is thrown overboard, the dinghy therefore displaces an extra amount of water with the mass 15 kg and the volume 15 dm^3. (The density of water is 1 kg/dm^3.) If the anchor is made of iron, it has a volume of about 2 dm^3. This is also the volume displaced by the anchor when it rests on the bottom of the pool. The difference, (15 – 2) dm^3 = 13 dm^3, gives rise to the lowering of the water level. Even in a very small pool with an area of 13 m^2 (the size of a bedroom), the level would change by only 1 mm, which is the same change as if we took out 13 dm^3 (3.4 U.S. gallons or 23 British pints) of water from the pool.

How physicists think. Our discussion started with some extreme assumptions about densities. Considering the fact that densities of real materials are not higher than about 23 000 kg/m^3 (the elements osmium and iridium), that is, 23 times the density of water, one may wonder whether it is good physics thinking to assume an almost infinite density. The answer is yes. In our "thought experiment" we are free to imagine any density. Thought experiments, which are sometimes referred to by the German word *Gedankenexperiment,* have played an important role in the development of physics. Two famous examples are the twin paradox in relativity theory and Schrödinger's cat in quantum mechanics. But there are also cases when it is not physically meaningful to make extreme idealizations (see problem 1.9 about a charged capacitor).

Additional challenge. Suppose that the dinghy is loaded with as many people as it can take. Then they all jump overboard and begin to swim. What happens to the water level in the pool then? And what happens to the level if they all drink some water from the pool? What happens to the water level in the pool if the dinghy starts to leak and sinks deeper but still floats? (See the solution at the end of this chapter.)

1.2 Ice in Water

The water does not spill over.

This problem has been around in popular books for more than a century. According to Archimedes' principle a floating object displaces water that has a weight equal to that of the object. We can imagine that the floating ice cube has "dug a hole" at the surface of the water. The hole is precisely large enough to hold water with the same mass as the ice cube. Therefore the level of the water in the glass will not change when the ice melts. A related, and more serious, problem deals with global warming. If ice resting on land melts, it will increase the level of the oceans, whereas the melting of floating ice will leave the level unchanged. Satellite observations have shown that, averaged over the Earth, the ocean level increases by about 3 mm (1/8 inch) per year. This is caused by global warming, but mainly because of the thermal expansion of the ocean water rather than the melting of ice.

Additional challenge. What is the answer to the problem if the ice cubes contain large bubbles of air, and if they contain grains of sand but still float? (See the solution at the end of this chapter.)

1.3 Accident in Aqueduct

The force on the supports increases by the weight of the tractor minus the weight of the water that the tractor displaces when it rests on the bottom of the aqueduct.

When the barge loaded with a tractor passes through the aqueduct, it displaces water that has the mass six tons (Archimedes' principle). A negligible fraction of that displaced water remains in the aqueduct. The rest flows into the much larger water reservoirs in the canal system, of which the aqueduct is just a small part, so that the water level in the aqueduct is unchanged. Therefore the force on the supports is unchanged when a barge passes. Six tons of water in the aqueduct is replaced by a barge and its load, which also have the mass of six tons.

But the forces on the supports are changed if the tractor rolls overboard. When it lies on the bottom of the aqueduct, the extra load is the weight of the tractor minus the buoyancy force associated with the water displaced by the tractor. Whether the barge itself remains in the aqueduct or not after the accident is irrelevant for the forces on the supports.

Any caveat? The most common version of this problem is to ask what is the added load on the supports when a barge enters the aqueduct, with the well-known answer that the load does not change at all. Here follows an aspect of the problem that does not have the slightest practical consequence, but still is of interest as a matter of principle. Suppose that the barge is loaded at a harbor, which can be far from the aqueduct. As the barge is loaded it sinks deeper and deeper into the water, and the water level rises everywhere in the system of directly connected waterways. In particular, the level increases also in the aqueduct, although to an exceedingly small amount, of course. In principle the barge has caused a miniscule increase in the load on the supports. From now on, the load on the supports is independent of where the barge is. (Would this description change if there were locks in the canal?)

A much larger effect than what was discussed in the preceding paragraph can come from dynamic effects, for instance, waves created by the barge and pressure variations of the type described by *Bernoulli's equation* for fluids, as the barge passes through the aqueduct. But it would still be a small change in comparison with the weight of the barge or the tractor.

1.4 Floating Candle

The candle continues to burn until it is very short, and certainly for more than one hour.

A good first step is to simplify the problem and consider a special case. Ignore the nail, and let the candle be fixed in a candlestick rather than floating in water. After two hours its length is only 4 cm, according to the text in the problem. If that short candle is placed in the glass of water, it would still float with 1/8th of its length (now 0.5 cm) above the water level (Archimedes' principle). The trend is clear. The candle can continue to float, with less and less height above the water, and it burns until it is very short.

There is more to it, however. When the upper edge of the candle is very close to the water, the stearin there is cooled by the water and does not melt. Instead, as the flame burns, the depression around the wick deepens. (This is shown in the inset in fig. 1.1.) The candle becomes lighter, and it floats a little higher. Then stearin can melt at the rim again. The candle sinks, and so on. Of course this is a continuous, rather than a stepwise, process. The result is that the candle floats with the flame at an almost constant height above the water.

The candle floating in water is not just an interesting scientific puzzle but has been of practical importance. Before gas lamps or electric light was available in laboratories, many experiments relied on candles as light sources. With a candle in an ordinary candlestick, the level of the light source varies significantly as the candle burns. The floating candle solved this problem.

Any caveat? If the iron nail is too heavy, the average density of the nail plus the burning candle will eventually be higher than of water. In practice one could circumvent this problem by keeping the candle in a narrow tube filled with water, as is done with some decorative candlesticks.

1.5 Running in the Rain

It is often concluded that it is best to run, but as we shall see the problem is very complex and sometimes there is no definite answer.

Figure 1.6 makes the standard solution obvious. Assume that your body can be modeled as a thin sheet, of width D and height H. For the moment we neglect the horizontal parts like the head and shoulders. To formulate a mathematical model we assume that the vertical velocity component of the raindrops is u. Furthermore, we assume that there is a steady headwind with a velocity component w toward you. The raindrops are dragged along by the wind and have the same horizontal speed w. Your own speed relative to the ground is v. After the time t you have covered the distance $s = vt$. Now look at figure 1.6. All those raindrops that happen to be in the shaded region when you start running from the point A at $t = 0$ will hit your "front" before you arrive to the point B. One can say that you sweep through this region and collect all raindrops it contained at the moment you started. The volume of the shaded region is

$$DH(vt + wt) = DHs(1 + w/v)$$

The prefactor DHs is a constant that only depends on your own size (front area DH) and on the distance s from your starting point to your goal. We note that the vertical speed u of the raindrops does not enter the result. The amount of rain hitting your front has its smallest value (for $w > 0$) when your speed v is as large as possible. Therefore you should run!

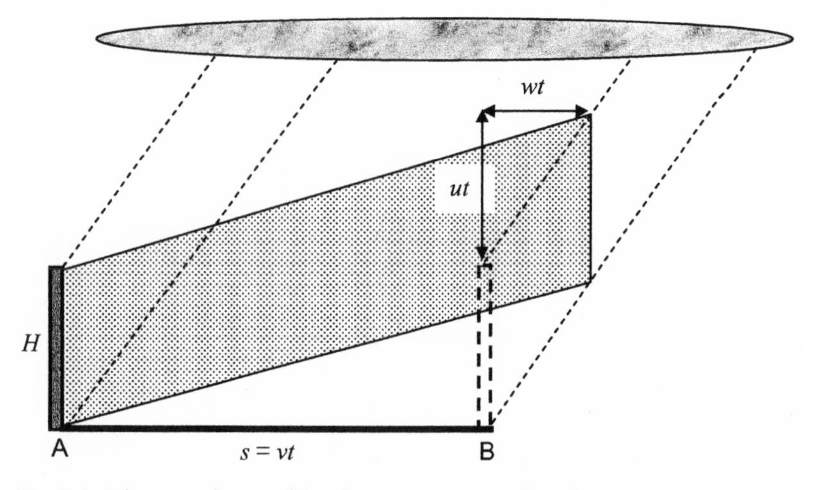

Fig. 1.6. Schematic figure of rain hitting a person of height H who takes the time t to move from A to B with speed v

A tailwind corresponds to a negative w. If your own speed relative to the ground is the same as the tailwind, that is, $w/v = -1$, the expression above suggests that you will not get wet at all. This is natural since each raindrop moves horizontally with the same speed as you have. You will never catch up with the drops in front of you, and the drops behind will not catch up with you. If there is no wind, the term w/v is zero, and it makes no difference whether you run. (But note the comment at the end of the solution to this problem.)

Our model assumed that a person is like an infinitely thin sheet and did not take into account the rain that hits the horizontal parts, for example, the head and shoulders. But in that case it should also be best to be exposed to the rain as short time as possible. We will quantify the effect at the end of the next section.

Any caveat? There are several aspects that deserve a closer analysis, or at least a comment.

We have argued that you sweep through a certain volume and are hit by all the raindrops that are in this volume at a given instant. The raindrops are not stationary but fall with a certain speed u. Does that matter? The answer is no. In every small volume in the air, the raindrops leaving that volume are continuously replaced by other raindrops, which fall into the same volume. The number of raindrops per volume of air is constant in time, as long as the intensity of the rain stays the same. In our model, the water hitting the front of the person does not depend on whether the raindrops fall fast or slowly, or if we just ignore their fall.

One may still find it surprising that it does not matter how fast the raindrops fall. It's worth considering this point in some detail. We certainly expect to get wetter in a heavy rain, and a heavy rain has large drops, which fall fast. The usual way of quantifying a rainfall is to give the *depth* per time of an assumed even layer of water on the ground (expressed, e.g., as the depth in millimeters, or inches, per hour) or as the *volume* per area and time (e.g., liters per square meter and hour; the symbol for liter is L or l). We shall call this the *intensity* I of the rainfall. In fact, expressing I as mm/h or as $L/(m^2 \cdot h)$ gives the same value. One liter is 0.001 m^3. Then, for instance, $7\ L/(m^2 \cdot h) = 0.007\ m^3/(m^2 \cdot h) = 0.007\ m/h = 7\ mm/h$.

In our case it is better first to introduce a quantity ρ that gives the volume of water per volume of air, for example, milliliters of water per cubic meter of air. The water that falls on the area A during the time t comes from a volume Aut in the air. The volume of that water is $(Aut)\rho$. Here u is the vertical speed of the drops. For the intensity (volume per area and time) we get

$$I = \frac{(Aut)\rho}{At} = u\rho$$

Compare this with the volume q of water that hits our front when we run in the rain. It comes from all the rain in a volume DHs of air (shaded region in fig. 1.6, assuming vertical rain). We get

$$q = (DHs)\rho$$

Combining the last two equations we can write the volume of the water that hits our front as

$$q = (DHs)\frac{I}{u}$$

Raindrops very quickly acquire the so-called terminal speed (terminal velocity), for which the gravitational force on the drops is balanced by the air resistance. The drops then fall with a constant speed, which increases with the drop size. For a given size we can assume that the speed u is a constant in the formula above for q. Then we have the expected result that how wet we get (expressed as q) is proportional to the intensity I of the rain. It might first seem that if u is very large, we will not get wet at all because according to our formula q tends to zero in that limit. This is misleading. What matters for us is the ratio

$$\frac{I}{u} = \rho$$

The speed u could also be very small, like in mist where the drops seem to hover in the air, but still give a constant ρ. Then the amount of water hitting the ground, and expressed by the intensity I, is very small. When we run through the mist, our front collects all the mist drops that are in our way.

Next we take a look at the horizontal parts of the body. Let their area be A_h. If we are exposed to the rain during a certain time *t*, we are hit by raindrops that would give a water layer of thickness *It* and volume $q_h = A_h It$. This volume can be expressed in quantities introduced above as

$$q_h = A_h u \rho t = A_h u \rho \left(\frac{s}{v} \right)$$

The ratio of the amount of water hitting head and shoulders to the water hitting our front is

$$\frac{q_h}{q} = \frac{A_h}{A_s} \cdot \frac{s}{H} \cdot \frac{u}{v}$$

Here $A_s = Ds$ is the area on the ground that we cover if our body width is *D* and we run the distance *s*. The first two ratios on the right-hand side do not depend on the speed *u* of the raindrops or on our own speed *v*, and the last factor is the ratio u/v of these two quantities. We can minimize the rain hitting us from above by running as fast as possible. This is obvious, since we are then exposed a shorter time to the rain from above. But does it matter? The speed of falling raindrops typically is $u = 6$ m/s. A walk has a speed of about $v = 1.5$ m/s and running (perhaps carrying a briefcase) could typically be at the speed $v = 3$ m/s. Then running, rather than walking, would half the ratio q_h/q. Is it worth it, considering that the rain *q*, which hits our front, is independent of whether we walk or run? We could also try to shield ourselves from above, for instance with a newspaper. And is our model account of the rain falling on the horizontal parts accurate enough? If we walk, our arms and legs can be held in almost straight and vertical positions, but if we run they will be bent and therefore present a larger horizontal area. Furthermore, leaning forward decreases our front area, but at the same time the horizontal area increases. Is that good tactics?

Finally, a comment on our statement that in a tailwind we could avoid getting wet, if we run as fast as the horizontal wind speed. This can only be partly true, because our body will disturb the wind pattern. In fact, the flow of air around the body will always be affected to some extent when we run, even with no wind. It is less important in heavy rain because of the inertia of the large raindrops.

We conclude that the popular problem of whether one should walk or run in the rain does not always have a definite answer. There are good arguments in favor of running, but in reality the advantage may be small and not worth the effort, or even turned to a slight disadvantage.

1.6 Reaching Out

A minimum of four books is needed.

The center of gravity of the combined pile must always fall inside the "support area," that is, the area on the table where the bottom book rests. If you try to analyze the problem by putting one book after another on top of each other, it becomes a bit complicated to calculate the maximum overhang. It is much easier to add the new book "from below." The idea is to add each book with its edge exactly below the center of gravity of the pile of books above.

We first consider a stack of two books, with a third book added from below, as in figure 1.7. It is added so that the edge of the lowest book is vertically below the combined center of gravity of the two books above.

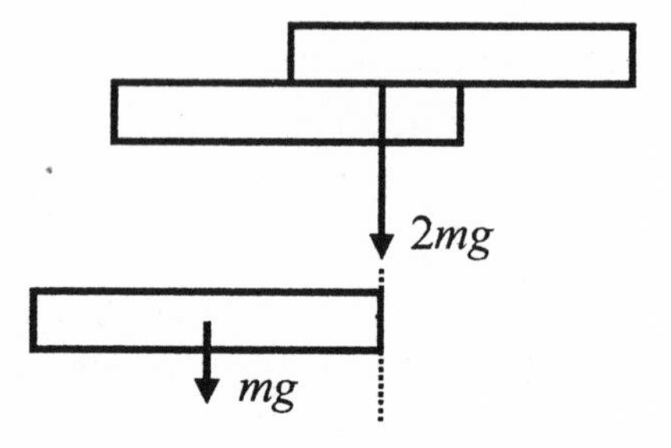

Fig. 1.7. Adding one book from below to a stack of two books

The analysis for an arbitrary number of books can be given a mathematical treatment as follows. Let the mass of a single book be m and its "height" (horizontal length in the stack) be $2h$. Choose the origin, $x = 0$, of the overhang coordinate x to be vertically below the center of gravity of a pile of r books. Put this pile on top of a book, with $x = 0$ at the edge of that book. The pile with $r + 1$ books has its combined center of gravity at $-x_{r+1} = -h/(r+1)$, as is obvious from the schematic illustration (fig. 1.8) and the equilibrium condition

$$rmgx_{r+1} = (h - x_{r+1})mg$$

with the solution

$$x_{r+1} = \frac{h}{r+1}$$

This is the additional overhang we get each time a book is added to the pile.

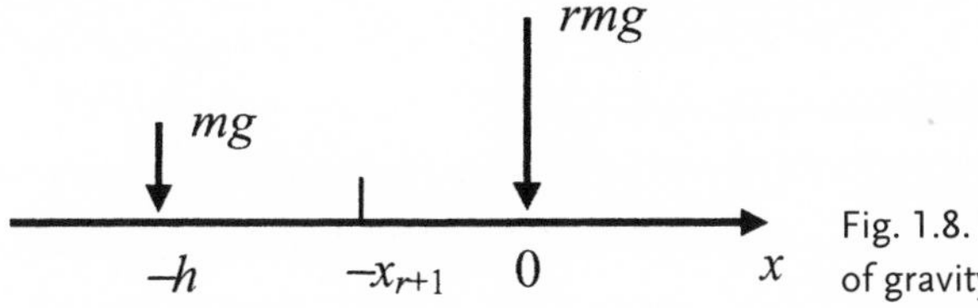

Fig. 1.8. Finding the center of gravity of $r + 1$ books

Following this scheme we see that with one book ($r = 0$), the maximum overhang is $x_{r+1} = x_1 = h$, with two books ($r = 1$) it is $x_1 + x_2 = h + h/2$, with three books $x_1 + x_2 + x_3 = h + h/2 + h/3$, and so on. This sequence is the well-known divergent harmonic series. For very large r the sum is approximately $h \ln r$. Thus, ignoring practical complications, the overhang could be made arbitrarily large. Table 1.1 gives the first six terms in the harmonic series, with overhang in units of h and with three significant digits. We see that a minimum of four books is needed for the top book to clear the edge of the table (overhang > $2h$). In the hypothetical case of 272 400 600 books, the overhang would be 10 books ($20h$).

Table 1.1. Maximum overhang, in units of half a book height

Number of books	Overhang
1	1.00
2	1.50
3	1.83
4	2.08
5	2.28
6	2.45

If you want to try this yourself, you may not have four identical books easily available, but coins make an excellent substitute. Instead of using books or coins you may also try to stack ordinary playing cards so that they reach far out from the edge of a table.

How physicists think. A creative mind may ask if it is possible to get a larger overhang if the books (coins, et cetera) are stacked in a different way. In the geometry in figure 1.9, one precisely reaches the desired overhang with only three books. Some readers may object that this solution breaks the rules of the puzzle, but "breaking the rules" is often the key to "breakthroughs." If you accept the breaking of rules you may try to get *two* playing cards to clear the edge of the table, with only *three* cards in total. It can be done in a very simple way, and without any damage to the cards!

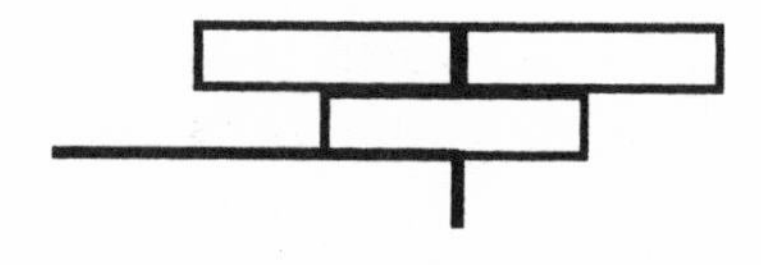

Fig. 1.9. In this stack of only three books, one book may precisely clear the edge of the table.

Any caveats? There is a difference between books and coins, because books can be somewhat compressible. In the lower parts of the pile they deform to a slight wedge shape. It might be a challenge for your dexterity to reach an overhang by one full book size with only four books.

Additional challenge. Three bricks, all 24 cm long, lie near the edge of a table, as shown in figure 1.10. You push on the lower left brick until one brick falls from the table. What is the minimum overhang relative to the edge of the table when this can happen? (See the solution at the end of this chapter.)

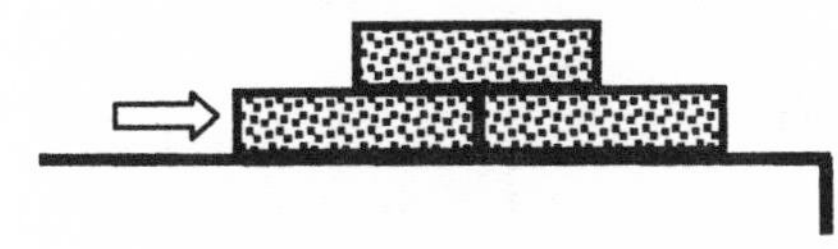

Fig. 1.10. When does a brick fall over the edge of the table?

1.7 Resistor Cube

The largest resistance between two corners is 10 Ω.

We first consider the resistance between the two most distant corners, for example, A and H in the figure. If a potential is applied between A and H, the corners B, C, and D will all be at the same potential, for symmetry reasons. Therefore we can connect B, C, and D and get three parallel resistors of 12 Ω between A and (B, C, D). The resulting resistance is 4 Ω. In the same way, the corners E, F, and G can be connected. The resistance between H and (E, F, G) is also 4 Ω. Finally, we see that (B, C, D) is connected to (E, F, G) by six resistors of 12 Ω in parallel, giving an effective resistance 2 Ω. In total we get $4\ \Omega + 2\ \Omega + 4\ \Omega = 10\ \Omega$ between the corners A and H.

The problem asked for the largest resistance between *any* corners. It remains to show that the resistance between two more close corners is less than 10 Ω. Although the result is intuitively clear, it is instructive to go through the detailed arguments. If a potential is applied between A and E, the symmetry of the system implies that the corners (B,C) are at the same potential as the corners (F,G). Then there is no potential difference between G and C, or between F and B, and the links G-C and F-B can be taken out. We now have a network where the resulting resistance 12 Ω for the link A-(B,C)-E is in parallel with the 36 Ω of the link A-D-(G,F)-H-E. The total resistance between A and E thus is exactly 9 Ω.

Finally, we consider the resistance between the two nearest-neighbor corners A and B. This is a more complicated problem to solve exactly, but as the original problem was posed, we only need to show that the resistance between A and B is *less* than 10 Ω. Remove resistors so that only the network in figure 1.11 remains, with 12 Ω for A-B in parallel with 36 Ω for A-D-F-B, which gives 9 Ω between A and B. When resistors are removed the total resistance between any two points in a network must increase (or remain unchanged) because we are taking away some paths for the current. The resistance between A and B in the full cubic network cannot be larger than 9 Ω.

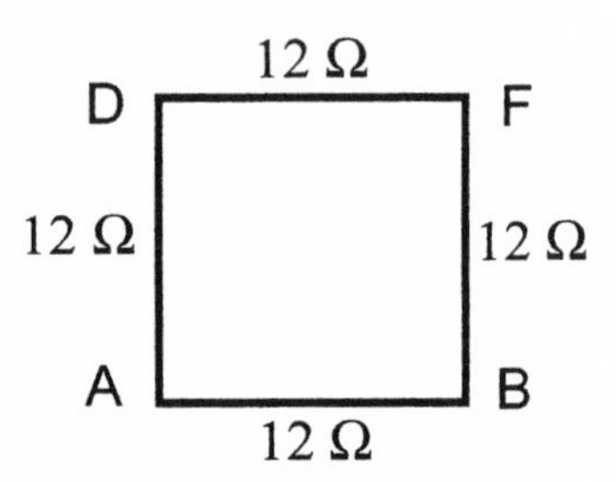

Fig. 1.11. The resistance between A and B is greater for this network than for the cubic network.

The resistance between two nodes in a network can be used to define a *metric,* that is, a mathematical measure of "distances" in graphs. Let a graph (finite or infinite) consist of links that are all assigned unit resistance. The distance between nodes i and j in the graph is defined as the effective resistance between i and j (as measured if a battery is attached to i and j). Table 1.2 gives the resistances R between the nodes in the five Platonic bodies when the resistance of a single link is R_0. The distances are then obtained as $D = R/R_0$. We see that the resistance between two nearest neighbors in our resistor cube, which was found above to be smaller than 9 Ω, is in fact 7 Ω. In an infinite square lattice, the distance between nearest-neighbor nodes is $\frac{1}{2}$. Between next-nearest neighbors it is $2/\pi \approx 0.64$.

Table 1.2. The resistance R between nodes in graphs defined by the five Platonic bodies when the resistance of a single link is R_0. The distance between two nodes can be defined as $D = R/R_0$.

Network	R_0	R
Tetrahedron	2	1
Cube	12	7, 9, 10
Octahedron	12	5, 6
Icosahedron	30	11, 14, 15
Dodecahedron	30	19, 27, 32, 34, 35

How physicists think. The solution above illustrates several useful methods in the analysis of networks:

- Points having the same potential can be connected.
- A resistor connecting points with the same potential can be deleted.
- *Deleting* a resistor between two nodes cannot *decrease* the resistance between any points in the network.
- *Adding* a resistor between two nodes cannot *increase* the resistance between any points in the network.

Additional challenge. Use the preceding ideas to solve the following problems. (See the solutions at the end of this chapter.)

a. Six resistors with resistance 6 Ω form the sides of a tetrahedron. What is the resistance between two corners?
b. Six resistors, with resistances 1, 2, 3, 4, 5, and 6 ohms form the sides of a tetrahedron. Show that the smallest possible resistance between two corners is less than 0.763 Ω.

1.8 One, Two, Three, Infinity

The resistance is 2.73 kΩ (or 2.7 kΩ).

Rather than immediately trying to solve the problem with seven steps, it is a good idea first to consider ladders with only one or two steps. They are easily found to have resistances

$$R_1 = 3R, \qquad R_2 = (2 + 3/4)R = 2.75R$$

where R is the resistance of a single resistor. The ladder with three steps is only slightly more complicated. We get

$$R_3 = (2 + 11/15)R \approx 2.733R$$

For each step that is added, the total resistance gets smaller, but at a rapidly decreasing rate. It seems clear that if we only want a result with an error less than 1 %, we can stop here. In practice, the resistance of a ladder with three or more steps is the same as that of an infinite ladder.

To make that point even more clear, we derive a recursion relation that relates the resistance R_{N+1} of a ladder with $N+1$ steps to that of a ladder with N steps. The ladder with $N + 1$ steps is constructed by adding one "step unit" to *the front end* of a ladder with N steps (rather than adding it to the far end of the ladder; see fig. 1.12). We get

$$R_{N+1} = R + \frac{RR_N}{R+R_N} + R = \frac{2R^2 + 3RR_N}{R+R_N}$$

The resistance converges very rapidly as N increases. In the limit $N \rightarrow \infty$ we have $R_{N+1} = R_N = R_\infty$. The recursion relation gets the form

$$R_\infty(R + R_\infty) = 2R^2 + 3RR_\infty$$

or

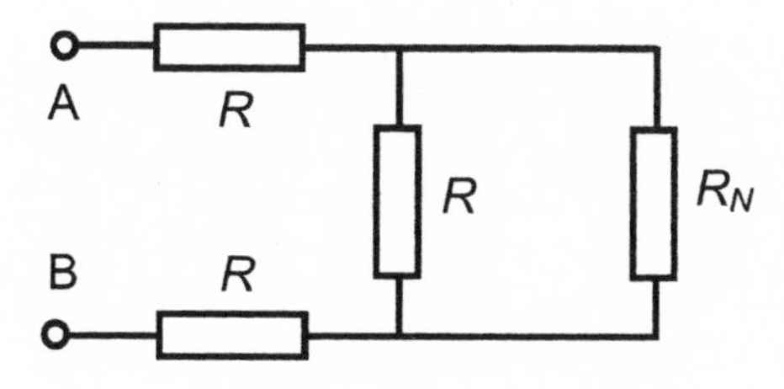

Fig. 1.12. A resistor ladder with $N + 1$ steps is formed by adding one "step unit" (three resistors) to a ladder with N steps.

$$R_\infty^2 - 2RR_\infty = 2R^2$$

This quadratic equation has the solution

$$R_\infty = R(1+\sqrt{3})$$

How physicists think. In a real problem, with commercial resistors labeled 1 kΩ, we may expect the resistance to show variations of at least 1% and perhaps much more. The small cylindrical resistors that previously were common in electronic equipment have colored rings, which indicate the resistance in ohms with two or three significant figures. There is also a ring giving the multiplier in powers of ten, and another ring giving the tolerance in percent. As is seen from table 1.3, it would require an unreasonably accurate value of R to give rise to any difference between the resistance of an infinite ladder and of a ladder with seven steps. The eight first digits in the resistance of a ladder with seven steps are the same as for an infinite ladder.

It is of interest to investigate the convergence rate of the recursion relation. Figure 1.13 shows $\log_{10}(D)$ as a function of N, for $N =$ 1 to 9, where D measures the approach of R_N to R_∞,

$$D = (R_N - R_\infty)/\text{ohm}$$

The graph is very close to a straight line in a log-lin plot, that is, it shows an exponential convergence of R_N with increasing N. The ex-

Table 1.3. The resistance R_N of ladder with N steps (in kΩ), when each resistor in the ladder has the resistance 1 kΩ

N	R_N
1	3.00000000000000
2	2.75000000000000
3	2.73333333333333
4	2.73214285714286
5	2.73205741626794
6	2.73205128205128
7	2.73205084163518
8	2.73205081001473
9	2.73205080774448
10	2.73205080758148
11	2.73205080756978
12	2.73205080756894
13	2.73205080756888
∞	2.73205080756888

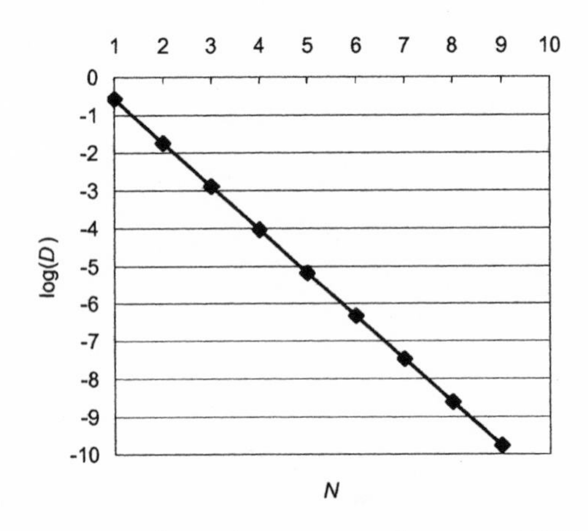

Fig. 1.13. $\log_{10}[(R_N - R_\infty)/\text{ohm}] = \log_{10}(D)$ as a function of N, for $N = 1$ to 9.

pression “One, two, three, infinity” gives a good characterization of our resistor ladder.

1.9 Lost Energy

The conventional answer is that the missing energy becomes heat in the connecting wires, because they cannot have a resistance that is precisely zero. This answer is not quite satisfactory. There are many ways of dissipating the missing energy. It can be converted to joule heat in the connecting wires, but it can also be radiated out, like electromagnetic waves from an antenna. Sparks created when the switches are closed give rise to heat, sound, light, and chemical reactions. The relative importance of various loss mechanisms depends on details in the circuit and on how it is closed. Dissipation by electromagnetic radiation can take place even in a model circuit where resistive or inductive components are strictly absent. Those electromagnetic waves arise in the connecting wires as well as in the capacitors themselves. Thus one cannot say exactly where the energy has gone. In any model that allows for at least one nonideal feature of the system, one can account for all the “missing” energy. A final remark: Does anything significant happen when the first of the two switches is closed?

How physicists think. The problem is well known. It is often said to exemplify a paradox, but the paradox arises from the unphysical idealized assumptions. The laws of Nature, for instance, the conservation of energy and momentum and also the thermodynamic relations derived from basic laws are *always* true. In a description of an actual physical phenomenon, we make use of a *model*. That means a simplification—certain effects are just ignored, while others are included. For instance, one would not try to include the Earth’s magnetic field in a model that describes the damage when two cars collide on a highway. But in some cases a model may leave out aspects that cannot be ignored. Our problem with the two capacitors is an illustration of that.

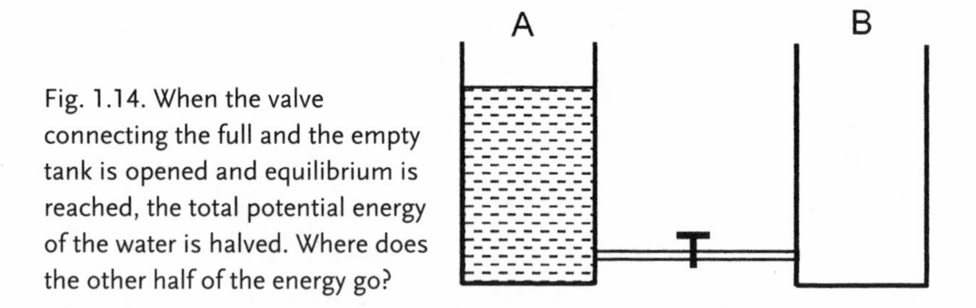

Fig. 1.14. When the valve connecting the full and the empty tank is opened and equilibrium is reached, the total potential energy of the water is halved. Where does the other half of the energy go?

The understanding of a physical phenomenon is often deepened if one can think of a similar phenomenon that is well known or is easier to analyze. Consider two equal tanks with water, connected by a tube and a valve as in figure 1.14. One tank is filled to the height h and the other is empty. The total mass of water is m. The potential energy E_A of the water in tank A (relative to the base of the tank) is

$$E_A = \frac{mgh}{2}$$

When the valve is opened, water flows into tank B until the level is the same in A and B. The height of water is $h/2$ and the mass is $m/2$, in each tank. The total potential energy is

$$E_A + E_B = \frac{m}{2} \cdot g \cdot \frac{h}{4} + \frac{m}{2} \cdot g \cdot \frac{h}{4} = \frac{mgh}{4}$$

Half of the energy is "lost." Eventually it will appear as heat, warming the water. If the valve is opened very rapidly (like in the rapid closing of switches in our capacitor problem), the water levels will oscillate somewhat as they approach the equilibrium. This obvious result for the water tanks suggests that one should consider the possibility of damped oscillations also in the capacitor case.

Another example of a paradox, which arises from the neglect of an important physical aspect, appears in the thermal conductivity λ of a gas. The result derived from kinetic gas theory says that λ does not depend on the gas pressure. That is surprising because in the limit that there is no gas at all, we certainly cannot have an unchanged thermal conductivity. The solution of the paradox is that the kinetic gas theory requires the mean free path ℓ of the gas particles to be much smaller than the characteristic dimensions L of the gas container. As the gas pressure is lowered, ℓ eventually becomes comparable to L. Then one enters a different physical regime for the energy transport by the gas.

1.10 Simple Timetable

The arrival time is always 42 minutes after the departure.

The train performs a harmonic oscillation, back and forth between the end points. With a geometry as in figure 1.15, the gravitational force on the train (mass m) along the tunnel is

$$F = F_{\mathrm{G}} \cos \alpha = F_{\mathrm{G}} \frac{x}{r}$$

Here F_{G} is the gravitational force in the direction of the center of the Earth;

$$F_{\mathrm{G}}(r) = \frac{GmM(r)}{r^2}$$

G is the constant of gravity and $M(r)$ is the mass of the Earth *inside* the radius r,

$$M(r) = \frac{4\pi}{3} \rho r^3$$

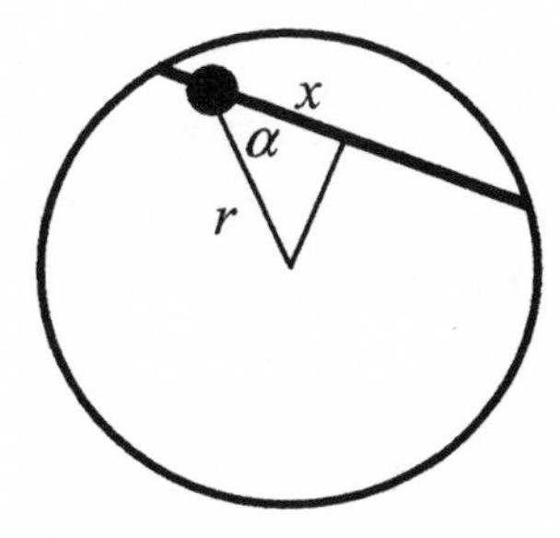

Fig. 1.15. The train (black dot) slides frictionless in a tunnel (fat line) through the Earth.

where ρ is the mass density of the Earth. Note that there is no net force on the train from the mass of the Earth that lies in the shell *outside* the radius r. This is a consequence of the $1/r$-type potential. An analogous result arises in electrostatic problems, for instance as in *Faraday's cage* where there is no electric field inside the cage.

Choose the x axis along the direction of the tunnel, with $x = 0$ at the center of the tunnel. The gravitational force that accelerates or retards the train is always towards $x = 0$. Newton's equation of motion, $ma = F$, now gives

$$m\frac{d^2x}{dt^2} = -m\frac{4\pi}{3}\rho G x$$

This is the equation of a harmonic oscillator with force constant

$$k = m\frac{4\pi}{3}\rho G$$

A well-known formula gives the angular frequency ω of a harmonic oscillator in terms of the force constant and the mass as

$$\omega = \sqrt{\frac{k}{m}} = \sqrt{\frac{4\pi\rho G}{3}}$$

The time T for *one half* oscillation, that is, from one turning point to the other, is

$$T = \frac{\pi}{\omega} = \sqrt{\frac{3\pi}{4\rho G}}$$

This is also the travel time from one capital to another in our problem. Suppose that we now want a numerical value for the time T, and there are no tables of physical constants readily available. We may not remember the values of ρ and G, but realize that the acceleration of gravity at the surface of the Earth, $g = 9.8\ \mathrm{m/s^2}$, can be written

$$g = \frac{GM(R)}{R^2} = \frac{4\pi\rho G}{3}R$$

where R is the radius of the Earth. Then

$$T = \sqrt{\frac{3\pi}{4\rho G}} = \pi\sqrt{\frac{R}{g}}$$

The value of R is easily obtained if we recall that the meter unit was once defined as $1/10^7$ of the distance from the Pole to the Equator, that is, $\pi R/2 = 10^7$ m. With $g = 9.8$ m/s^2 we get $T \approx 42$ min.

How physicists think. You may note that the transit time 42 minutes is half the time for a satellite to circle the Earth. Such a relation cannot be a mere coincidence. In both cases there is an object (a train or a satellite) that moves under the influence of the gravitational field of the Earth, and in paths whose length is conveniently expressed in the Earth's radius R. (The speed of the satellites is chosen such that their orbits lie close to the surface of the Earth, when seen on the scale of the Earth's radius.) A measure of the strength of the gravitation is given by the acceleration g. The quantities R (with SI unit m) and g (with SI unit m/s^2) can be combined in *one and only one* way to give a quantity with the dimension of time (SI unit s). This unique combination is

$$\sqrt{\frac{R}{g}} \approx 800 \text{ s}$$

Any problem that involves *only* R and g, and where we ask for a time, must have a result that can be written as a numerical factor multiplied by $(R/g)^{1/2}$. In our case that factor is π for the tunnel problem and 2π for the satellite orbit problem. But even the answer 800 s ≈ 13 min is of the right order of magnitude for both these problems. Such an "educated guess" can give a valuable check on a calculation. For instance, if the equation for T is used in a computer program, and the numerical value assigned to G (it should be 6.67×10^{-11} N m^2/kg^2) by mistake has been given a wrong power of ten, the error might be detected in a comparison with the *characteristic time* $(R/g)^{1/2}$, which is obtained in a simple dimensional analysis.

In the discussion in the previous paragraph we applied *dimensional analysis* based on the quantities R and g. But if we go back to the expression

$$T = \sqrt{\frac{3\pi}{4\rho G}} = \pi\sqrt{\frac{R}{g}}$$

we see from the first equality that T does not depend on the radius R of the Earth—only on the Earth's average density ρ and on a constant of nature, G. In principle we would get the same transit time through a sphere as small as a tennis ball, provided of course that the sliding particle is not affected by other gravitational forces. (Would it be sufficient to orient the tunnel in the ball horizontally on the Earth? If so, what accuracy would be needed?) With $\rho = 5520\ \mathrm{kg/m^3}$ and $G = 6.67 \times 10^{-11}\ \mathrm{m^3/(kg \cdot s^2)}$ we get the characteristic time

$$\sqrt{\frac{1}{\rho G}} \approx 1650\,\mathrm{s}$$

This is of the same order of magnitude as the value $(R/g)^{1/2} \approx 800$ s we obtained above. How could it be that we first get a time T that seems to *depend explicitly* on R and then find another result that is *independent* of R? The explanation is that the acceleration of gravity g at the Earth's surface also depends on R. The difference by about a factor of 2 between the two characteristic times, 800 s and 1650 s, is not unreasonable in arguments that only rest on a dimensional analysis.

Finally, we should remark that after all there is a fundamental difference between the tunnel and the satellite problems. The tunnel transit time depends on the *distribution* of mass in the Earth (which was assumed to be uniform in our calculation), whereas the satellite orbital period only depends on the *total* mass of the Earth.

Of course the problem with the tunnel train is completely unrealistic, but that did not prevent *Time* magazine from presenting it as science news (11 February 1966). In fact, the tunnel problem was known to Robert Hooke, who wrote about it in a letter to Isaac Newton almost 400 years ago. An 1883 article in the French journal *La Nature* had the title "De Paris à Rio-Janeiro en 42 minutes 11 secondes." In chapter 7 of the novel *Sylvie and Bruno Concluded* by Lewis Carroll (1893), a German expert ("Mein Herr") explains that in his country "each railway is in a long tunnel, perfectly straight; so of course the middle of it is nearer the center of the globe than the two ends; so every train runs half-way down-hill and that gives it force enough to run the other half up-hill."

ADDITIONAL CHALLENGES

1.1 Dinghy in Pool

The water level remains the same. Any object that can float displaces an amount of water determined by the weight of the object. It makes no difference if it floats as a load in a boat, or if it floats by itself in the water. If you swallow some water, the mass of your body increases. The corresponding increase in the volume displaced by your floating body (Archimedes' principle) exactly compensates for the loss of water in the pool.

1.2 Ice in Water

Whether the ice contains bubbles of air or not makes no difference in the answer to the problem. With grains of sand in the ice, the water level will decrease somewhat when the ice melts. That is because sand has a higher density than ice or water. The "hole" created at the surface of the water by the floating piece of ice is larger than the volume of the sand grains plus the melted ice. This is analogous to problem 1.1 about a dinghy in a pool.

1.6 Reaching Out

The bricks can have such rough surfaces that one does not know exactly where the top brick makes contact with the two others. Therefore we consider the special case in which a brick is about to fall over the edge and lifts up the center of the brick resting on it, as in figure

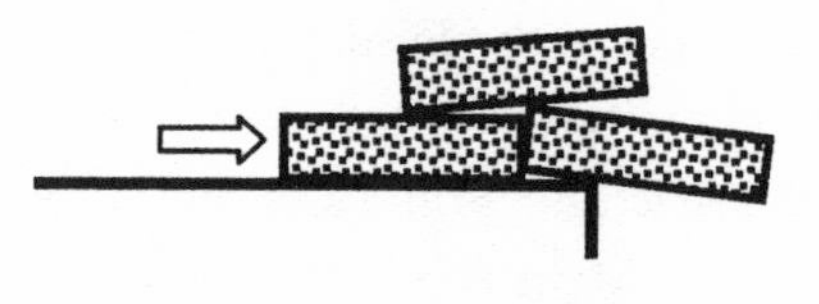

Fig. 1.16. Geometry when one brick starts to fall from the edge of the table

1.16. Force equilibrium requires that $\frac{3}{4}$ of the outermost brick is beyond the edge. The overhang thus is 18 cm. This situation represents an idealization. With friction taken into account, the overhang could be larger before the brick falls from the table.

1.7 Resistor Cube

a. Apply a potential between A and B in the equivalent flattened network in figure 1.17. The link CD can be deleted because C and D have the same potential. Then we have three resistors in parallel, with resistances 6 Ω, 12 Ω, and 12 Ω, giving the resulting resistance 3 Ω between A and B.
b. We are not asked to give an exact value for the effective resistance, but only to show that it is smaller than 0.763 Ω. Make the tetrahedron "flat" as in figure 1.17, and delete the dashed link CD. With resistance values (in ohm) AB = 1, AC = 2, CB = 3, AD = 4, and DB = 5, we get a parallel coupling of resistors with total resistances 1, 5, and 9. The total resistance between A and B then is 45/59 = 0.7627. When the remaining resistor (6 ohm) is placed between C and D we open a new path for the current, and the effective resistance between A and B is lowered. Therefore it is possible to connect the six resistors in a tetrahedron so that the smallest resistance between two corners is less than 0.763 Ω.

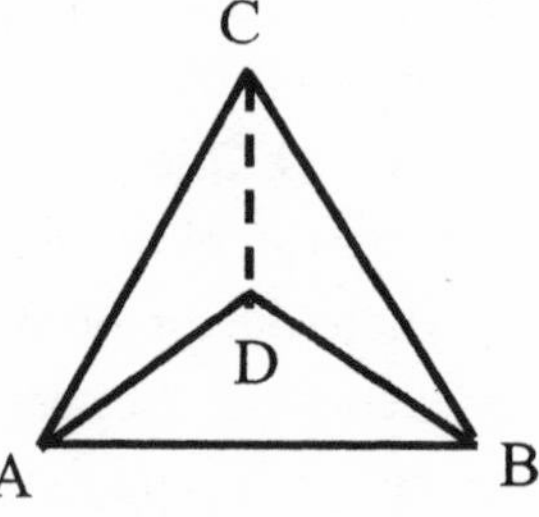

Fig. 1.17. Resistance between A and B is lowered if a link is introduced between C and D to form a tetrahedron of (unequal) resistors.

2 No Math Required

This chapter contains problems of increasing difficulty for which only qualitative answers are required, although simple mathematics may give additional insight.

PROBLEMS

2.1 Moving Backward?

You ride an ordinary bicycle, in ordinary gear. Are there any points on the bicycle that don't move forward relative to the road?

2.2 Heating Water

In 1895 the first large-scale hydroelectric power plant was installed at Niagara Falls, New York, and by 1900 there were ten generators, rated at 5000 horsepower each. Suppose that we let this electric energy feed a giant electric heater placed in an exit tube after the turbines. Would that energy suffice to bring the water in the exit tube to boiling, or would it increase the temperature by not more than about 1 K (2 °F)?

2.3 Bright Lamps?

A network has three equal lamps, connected as in figure 2.1. When switch 1 is closed, lamp A lights up with normal brightness. Are

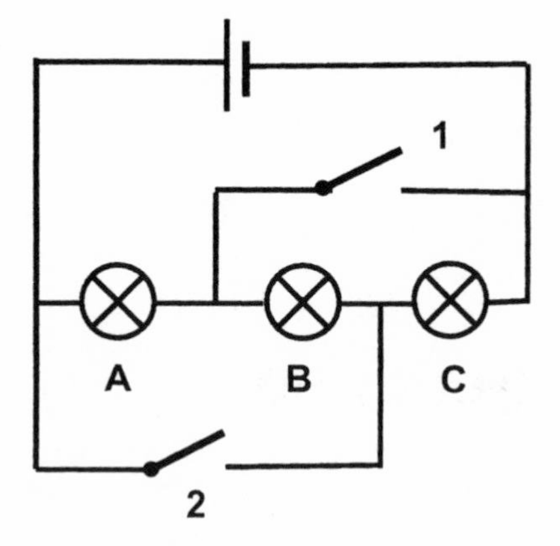

Fig. 2.1. Compare the brightness of the lamps when the switches are closed.

there any lamps that light with normal brightness if switch 2 is also closed?

2.4 Low Pressure

The distance covered by a bicycle can be measured with a meter connected to the front wheel. A cogwheel advances one step for each turn of the front wheel. Does the tire pressure affect the measured distance? This is not as innocent a problem as it might first seem. Many people with good scientific training have had heated arguments for either yes or no.

2.5 Site for Harbor

A country has a long, almost straight shoreline, where there are no natural sites for a harbor. One therefore decides to build only one harbor, from which straight railways go to two cities, A and B (fig. 2.2). Can you locate the position of the harbor, which minimizes the total length of the railways, by geometrically constructing straight lines on the map rather than by directly measuring distances on the map or doing mathematical calculations?

2.6 More Gas?

Can you fill your car's fuel tank with more or with less gasoline on a hot day? Because it is the chemical reaction of the fuel that pro-

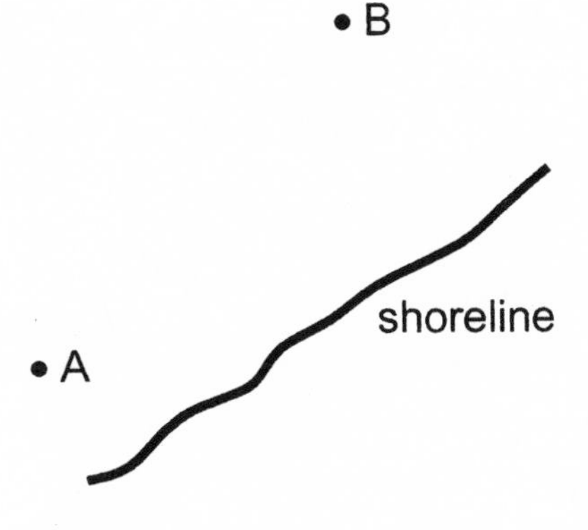

Fig. 2.2. Where should the harbor be located?

vides the source of energy, it is the number of fuel molecules in the tank that is of interest.

2.7 High Tension

You stand beneath an electric 400 kV power line with an ordinary magnetic compass in your hand. The power line runs in the north–south direction. In which direction does the red-painted part of the compass needle point?

2.8 Ocean Surface

Even in the absence of centrifugal effects from the Earth's rotation, the surface of the oceans would not have perfect spherical shape, because the mass distribution in the Earth does not have perfect spherical symmetry. Consider an underwater mountain, protruding from an otherwise almost flat seabed. Is the ocean surface above that mountain slightly depressed or does it form a small hump?

2.9 Mariotte's Bottle

An open tube T is inserted in a tight cork in a bottle, as figure 2.3 shows. There is an opening O in the side of the bottle. When the bottle is filled as in the figure, water flows out from the opening. The flow rate through the opening decreases gradually until it achieves a constant value q. Then it stays constant for a while, followed by a

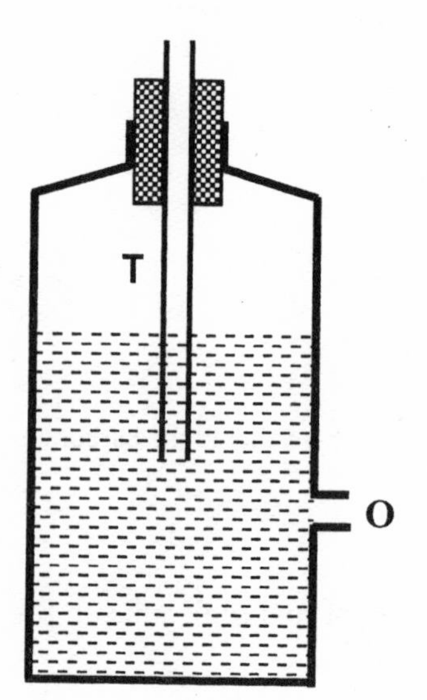

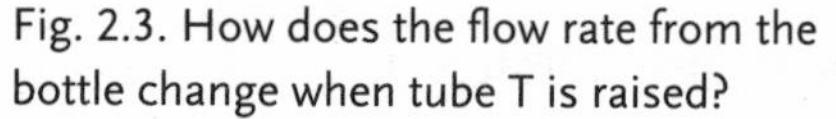
Fig. 2.3. How does the flow rate from the bottle change when tube T is raised?

decreasing flow rate, until the flow finally stops. Is the constant flow rate q just mentioned increased, decreased, or unaffected if the tube T is raised somewhat?

SOLUTIONS

2.1 Moving Backward?

Yes, the contact points between the tires and the road.

If you ride the bicycle with constant speed v, the axles of the wheels of course also move forward with that constant speed. Where the wheels touch the road, their speed relative to the road is zero (otherwise they would slip), while their diametrically opposite parts move forward with the speed $2v$ relative to the ground.

Any caveat? It is not quite true that the speed is zero at the contact point between the wheel and the road. The rubber in the tire deforms and there is a small slip so that, on the average, the rubber at the contact point moves slowly backward.

Could there be any other part that moves backward? Some bikes have an electric generator with a small wheel in contact with the side of the tire. The wheel rotates very rapidly, that is, it makes many turns per second. But no part moves backward relative to the road. If the axle of the generator wheel is vertical and is mounted at the top of the bicycle wheel, the side of the generator wheel that is in contact with the bicycle wheel moves forward with the speed v relative to the frame of the bicycle. The diametrically opposite point (at A) moves backward with the speed v relative to the frame (fig. 2.4). Since the frame itself moves forward with the speed v relative to the road, the point A has zero speed relative to the road. (To be more precise, the speed is not precisely zero because the generator wheel touches the tire a bit below the circumference of the tire.)

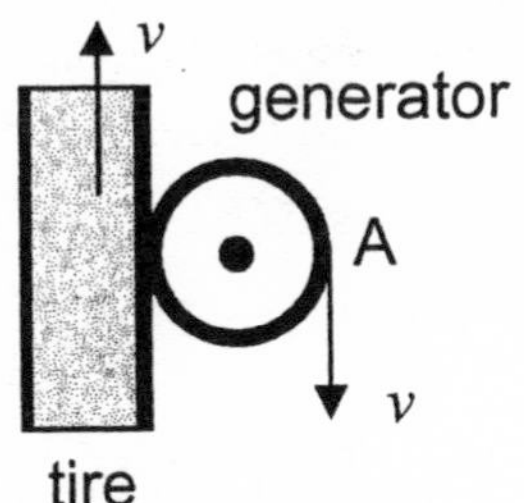

Fig. 2.4. Bicycle generator driven by the bicycle tire

Some people think that the pedal might move backward, when it is in its lowest position. Suppose that we could attach a chalk to the pedal and let the chalk trace out the pedal's trajectory on a vertical wall along the road. We describe the trajectory in a Cartesian coordinate system with x being the path of the bicycle frame and y being the vertical distance of the pedal (chalk) to the axis of the pedal system. Then,

$$x = vt - r \cos \omega t$$

$$y = r \sin \omega t$$

where v is the bicycle speed, t is the elapsed time, ω is the angular frequency of the pedaling motion, and r is the radius of the pedal crank (fig. 2.5). The relation between v and ω depends on the gear and the wheel radius R. During the time t, the pedals make $n = \omega t/(2\pi)$

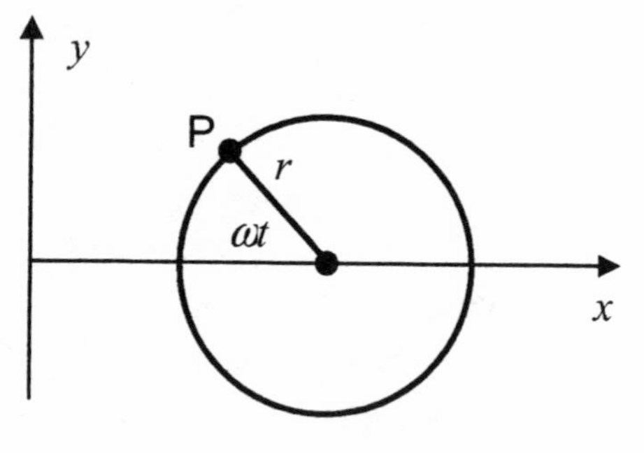

Fig. 2.5. Defining the coordinates (x,y) of the pedal P as the bicycle moves with speed v along the x axis

revolutions. With a gear G, the rear bicycle wheel makes nG revolutions and covers a distance $vt = 2\pi RnG$, that is,

$$x = 2\pi RG\frac{\omega t}{2\pi} - r\chi\cos t = r\left(G\omega t\frac{R}{r} - \chi\cos t\right)$$

The trajectory therefore has the same mathematical shape $y(x)$ as a function defined parametrically through the variable $u = \omega t$ ($a = GR/r$), that is,

$$x = au - \cos u$$

$$y = \sin u$$

This is called the *cycloid* function (fig. 2.6).

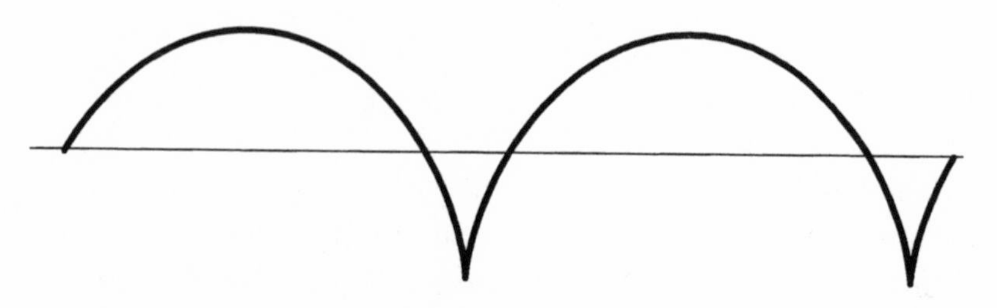

Fig. 2.6. The shape of the cycloid function

2.2 Heating Water

The increase in temperature is certainly much less than 1 K. It could never be larger than the temperature difference between the foot and the top of a free vertical waterfall, when only the change in the potential energy of the water is considered. If waterfalls with a scalding temperature at the foot existed, it would be a very well known fact.

In an ideal hydroelectric power plant, the potential energy of the water in the dam is completely converted to kinetic energy, and then to electrical energy in the turbines. Compare this with a vertical waterfall without any power plant. When the water hits the foot of the waterfall, part of the kinetic energy is converted to heat. Assume, as an extreme case, that *all* the potential energy is used to heat the water. Let water with mass m fall the vertical distance h. The available energy (to be converted to heat) is

$$Q = mgh$$

If the specific heat capacity of water is C, the temperature increase is

$$\Delta T = \frac{Q}{mC} = \frac{gh}{C}$$

With $C = 4.2\ \mathrm{kJ/(kg \cdot K)}$, $g = 9.8\ \mathrm{m/s^2}$, and an assumed height as large as $h = 100$ m (330 ft), we get the very modest temperature increase

$$\Delta T = 0.2\ \mathrm{K}\ (0.4\ ^\circ\mathrm{F})$$

Any caveat? Our result for the temperature increase is an overestimation, because it assumed an idealization in which all the potential energy is first converted to kinetic energy, $mgh = mv^2/2$, and then converted further to the heat Q without any losses. Although this solves our problem, one may ask how much of an overestimation it is. The water has a horizontal velocity u after it has hit the ground.

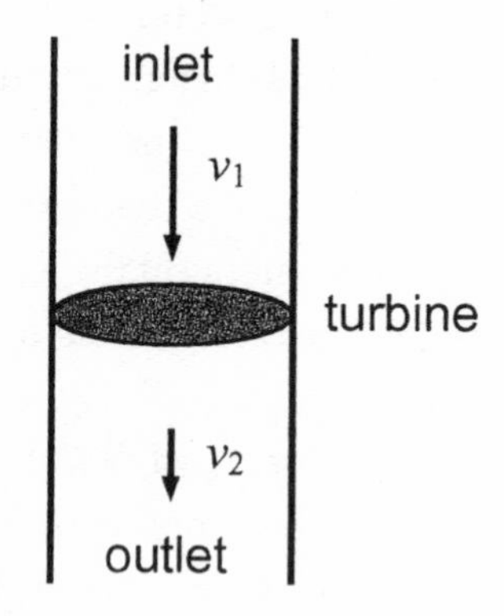

Fig. 2.7. Misleading schematic illustration of a hydroelectric power plant

But if $u << v$, this gives only a small correction to ΔT. Anyone who has watched the flow at the foot of a vertical waterfall knows that $u << v$, but what is it in a real hydroelectric power plant?

Consider figure 2.7. If the turbine completely converts the kinetic energy to electric energy, the flow velocity after the turbine must be zero. But the same amount of water that is fed through the tube into the turbine must also leave the tube after the turbine. If A_1 is the cross-section area and v_1 is the water speed in the tube before the turbine, with A_2 and v_2 describing the flow after the turbine, you must have

$$A_1 v_1 = A_2 v_2$$

This is a simple version of the *continuity equation.*

It is now obvious that A_2 must be much larger than A_1, so that we get $v_2 << v_1$. The figure, therefore, is very misleading. A real hydroelectric power plant can be constructed in such a way that more than 95 % of the potential energy of the water is converted to electrical energy. But you might have heard that at most 59 % of the kinetic energy in the wind can be used in a windmill. That is a theoretical limit, sometimes referred to as the *Betz theorem.* The idea behind the Betz theorem perhaps is best understood by using a water mill as an example. If the paddles in the wheel are moving with the same speed as the incoming water, they make no resistance to the flow and the generated power is zero. On the other hand, if the paddles are immobile, they can halt the water flow completely. But then no generator is rotating, and the power is again zero. The optimum energy conversion, 16/27 = 59 % of the power in the incoming flow, is obtained as a compromise between these two extremes. Obviously the turbine system in a hydroelectric power plant has a much more complicated geometry, so that almost all of the water's energy can be used.

How physicists think. Knowledge is in part to remember facts, but the *absence* of known facts also can contain information. In our example we noted that the temperature increase at the foot of the waterfall certainly could not reach the scalding temperature of 50 °C (about 120 °F). Such waterfalls would be so remarkable that everyone would have heard of them.

Here is another example relating to general knowledge. Readers of this book probably have encountered the question of whether one can boil eggs on the top of Mount Everest. The idea is that the boiling temperature of water depends on the atmospheric pressure. On the top of Mount Everest water boils at about 72 °C, or 162 °F. The egg proteins change their structure at about 63 to 66 °C, first in the white and then in the yolk. Theoretically, one may boil an egg on Mount Everest, but because the reaction rate of the proteins depends exponentially on the temperature, it would be an extremely slow process. Even at 85 °C (185 °F) it takes half an hour to get a tender egg. People living in mountainous regions, like Colorado in the United States, know that certain foodstuffs require a longer heating time than cooking requires at sea level. But it is never suggested that one should increase cooking times on a stormy day with low atmospheric pressure. If that had been necessary, everyone would know it. We can draw the conclusion that the more or less daily variations in the atmospheric pressure are (much) smaller than the difference in pressure between, say, Denver and New York. In Denver the pressure typically is 840 hPa (840 millibar) and in New York it is 1000 hPa, whereas the difference between a low-pressure and a high-pressure day is less than about 50 hPa.

There is an anecdote telling that when William Thomson (Lord Kelvin) was hiking in the Swiss Alps he found James Prescott Joule busy measuring the temperature in waterfalls, while Joule's young wife was waiting in their cab. If the story is true, Joule not only needed a very accurate thermometer to verify that the water is warmer at the foot of the waterfall. He might also have noted that the ambient air temperature typically decreases by 0.6 K (1 °F) for each increase in height by 100 m (300 ft). This difference is significantly larger than the maximum $\Delta T \approx 0.2$ K we calculated above for a 100-m-high waterfall. If Joule wanted to relate the temperature of the water to the change in potential energy he had to be careful to avoid indirect effects caused by variations in the atmospheric temperature.

2.3 Bright Lamps?

All three lamps have normal brightness when switches 1 and 2 are closed, because they are all coupled in parallel. This may be difficult to see at first, but here is a simple argument. Rather than focusing on the whole network we consider one lamp at a time. Each lamp has one side directly connected with one of the battery poles, and the other side of the lamp is directly connected with the other battery pole. This solution assumes that the internal resistance of the battery is so small that the voltage does not depend on whether one or three lamps are connected.

How physicists think. Drawing the right figure can give immediate insight in a physics problem. Here is another example (fig. 2.8). Find the resistance between A and B when the links in the network have the indicated resistances, in ohm. The problem becomes trivial if one realizes that the network to the left is identical with the network to the right. Apply a potential between A and B. The potential across

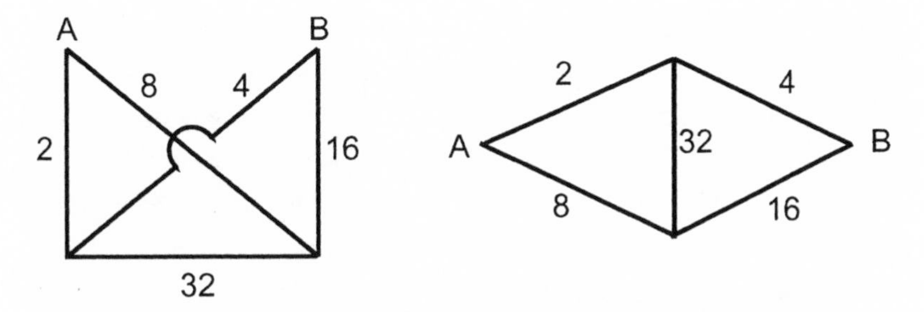

Fig. 2.8. Two seemingly different, but identical, networks

the 32-Ω resistor is zero because the network has the form of a balanced so called *Wheatstone bridge*. (The ratio 2:8 is the same as the ratio 4:16 for the resistances of the "arms" of the bridge.) We then have $(2 + 4)\ \Omega = 6\ \Omega$ in parallel with $(8 + 16)\ \Omega = 24\ \Omega$, which gives the total resistance $24/5\ \Omega = 4.8\ \Omega$ between A and B.

Additional challenge. A lamp and three identical resistors form a network as in figure 2.9. The applied potential is such that the lamp is shining brightly. A fourth identical resistor is added to the network in such a way that the lamp becomes dark. Where should this extra resistor be inserted? (See the solution at the end of this chapter.)

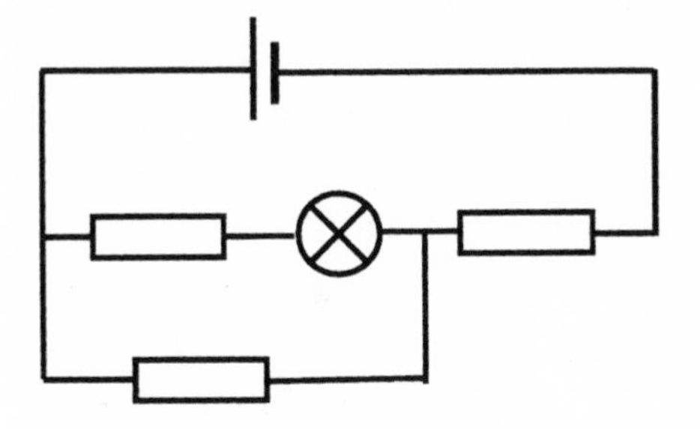

Fig. 2.9. Where should a fourth identical resistor be added, so that the bright lamp becomes dark?

2.4 Low Pressure

The recorded distance will be longer if the tire pressure is low.

The simple argument for this answer is that the effective radius of the front wheel (the distance from wheel axle to the point of contact with the road) is smaller if the tire pressure is low, so that the tire is more compressed. The wheel must make more turns for a given distance. This is also the correct description.

An (incorrect) argument that the recorded distance is unaffected by the tire pressure goes like this. The outer circumference of the tire has a certain length, which is almost independent of the pressure in the inner tube. When the wheel has made one turn, it has covered a distance on the road which is equal to that circumference. Now imagine that you cut the rubber tire to form a measuring tape, which is stretched out on the road. For each turn of the wheel, you have covered a distance on the road that is equal to the length of the measuring tape. Therefore it does not matter what the pressure in the tire is. However, this argument is misleading. The rubber in the tire is deformed and when the wheel rolls there is a slip between the tire and the road. (For this reason, the effective radius is not exactly equal to the distance between the wheel axes and the contact point on the road.)

How physicists think. Physics depends on experiments. Rather than getting caught in a theoretical dispute, one could try the following. Hold the bicycle upright with the valve of the front wheel located just above the contact point between the tire and the road. Walk with the bicycle along a straight line until the valve is again at its lowest position. Mark this point on the road. Then start over again in the same way, but now with very low tire pressure. The front wheel has made one full rotation before one reaches the mark on the road.

This problem may bring to mind an interesting observation regarding the odometer reading in cars. Odometers measure the number of rotations of a wheel. The outer tire radius is about 30 cm (1 ft). A worn tire has a radius that is decreased by almost 2 %, and the odometer reading increases by the same amount. Perhaps you regularly drive a long distance with the same car and the same tire pressure. The difference between the distance recorded by the odometer when the tires are new, and some years later when the tires are worn, should be observable. As a curiosity we note that a taxi driver was once fined because he had put on wheels that were too small to charge his customers for driving longer distances.

Additional challenge. A mother presents her daughter with a bicycle that has a measuring device as described in this problem. The proud daughter first rides the bike and measures a certain distance on the road. Then her mother rides the same bike and measures the same distance. Who will get the smallest value for the distance? (See the solution at the end of this chapter.)

2.5 Site for Harbor

The law of optical reflection can be used to locate the harbor. Mark where the reflection B′ of the city B is located in the "sea," if the shoreline is considered as a mirror. The harbor should be at H, where a straight line between A and B′ intersects the shoreline (fig. 2.10).

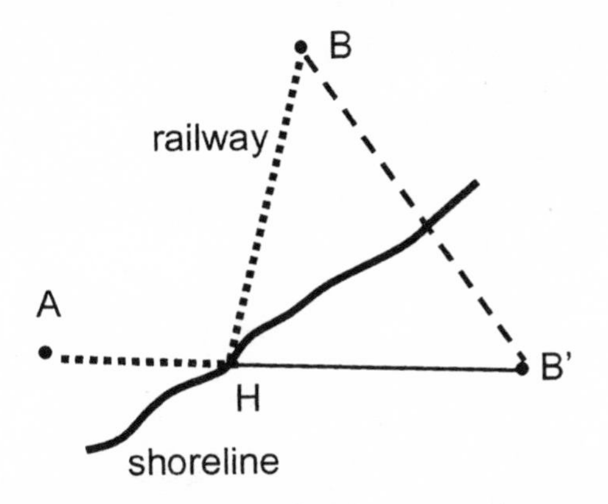

Fig. 2.10. Geometrical construction to find the point H that minimizes the distance AHB

How physicists think. The relation between mirrors and our problem is no coincidence. It expresses a very fundamental physical rule, known as *Fermat's principle,* for the propagation of light. Consider a light ray emitted from a point A and ending at a point B after reflection in a mirror. Of all different paths a, b, c, . . . between A and B, light follows that one, which minimizes the time it takes for the

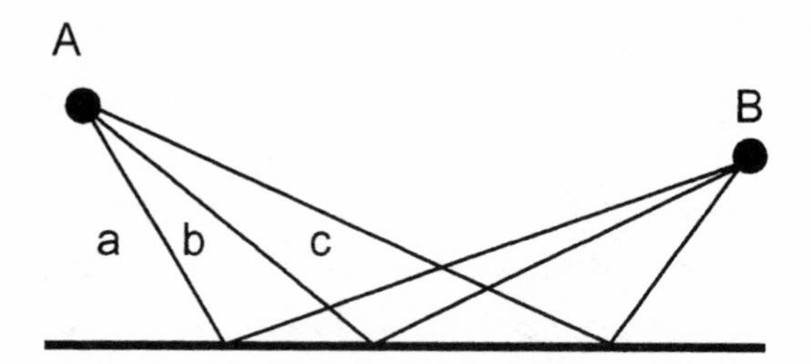

Fig. 2.11. A light ray takes the path from A to B that minimizes the total travel time.

light to travel from A to B (fig. 2.11). In this example the speed of light is constant. Minimizing the time, therefore, is the same as minimizing the distance traveled.

We can take the analogy between the railway and light propagation further. Suppose that the city A has an industry that requires many transports, so that the railway to A should have higher capacity and turns out to be 40 % more expensive than the railway to city B. Then the harbor should be located at a point H where, with notation as in figure 2.12,

$$\frac{\sin\alpha}{\sin\beta}=\frac{1}{1.40}$$

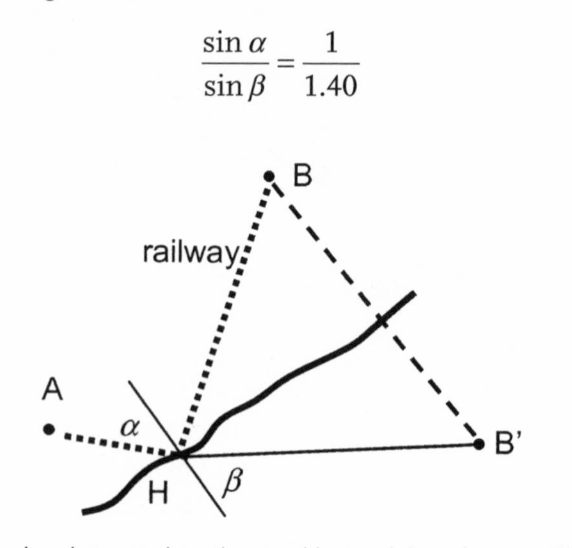

Fig. 2.12. Analogy between the railway problem and the refraction of light at an interface

Here we recognize *Snell's refraction law* for light. This example shows how thinking in physics can have applications in quite another field.

2.6 More Gas?

The tank can be filled with less gasoline on a hot day. The liquid expands significantly when the temperature increases from the rather low value in the underground storage tank to that of the ambient temperature. Therefore, if one fills the car's tank to the brim, it will flow over when the gasoline's temperature increases.

A first thought might be that the tank itself has a larger volume when it is hot, due to thermal expansion. This is true, but it is a very small effect. Let the tank be made of steel. The *linear* thermal expansion coefficient α of steel is about $12 \cdot 10^{-6}\ \mathrm{K}^{-1}$, and the *volume* thermal expansion coefficient β is three times larger, $\beta = 36 \cdot 10^{-6}\ \mathrm{K}^{-1}$. For a temperature difference $\Delta T = 20$ K (36 °F), the volume of the tank increases by a factor

$$1 + \beta\,\Delta T = 1.0007$$

Most liquids have a much larger thermal expansion than steel. For gasoline, $\beta \approx 900 \cdot 10^{-6}\ \mathrm{K}^{-1}$.

At a depth of about 3 m (10 ft) the temperature in the ground is almost constant throughout the year, and about equal to the annual average ambient temperature. We therefore let the temperature of the gasoline in the underground storage tank be 20 K lower than the ambient temperature of the hot day in our problem. The volume of the gasoline then increases by a factor

$$1 + \beta\,\Delta T = 1.018$$

This is 25 times greater than the change in the volume of the car's tank. Table 2.1 shows the volume changes with temperature for

Table 2.1. Temperature dependence of the volume of steel, gasoline, and water

	Volume at		
Substance	0 °C (32 °F)	15 °C (59 °F)	30 °C (86 °F)
Steel	1000	1000.5	1001.1
Gasoline	1000	1013	1027
Water	1000	1000.7	1004.2

steel, gasoline, and water, when the volume has been normalized to 1000 (in arbitrary units) at 0 °C (32 °F).

Any caveat? We considered the number of gasoline molecules, assuming that this is what matters when we are interested in the energy supply to the engine. But the efficiency of the combustion process, and the conversion from chemical to mechanical energy, can depend on the temperature in a very complicated way that varies from case to case. A simple scientific argument that focuses on only one aspect of a problem (here, thermal expansion) may give a very misleading answer when confronted with reality. In our example reality is so complex that one should rely on experiments rather than on a theoretical reasoning.

Water is well known to have a density maximum at 4 °C (39 °F), that is, the thermal expansion coefficient is strongly temperature dependent and is zero at 4 °C. One may wonder if such behavior is characteristic of most liquids, but water has very anomalous thermodynamic properties, contrary to gasoline. In table 2.1 a temperature-independent β for gasoline is assumed, although β in fact increases somewhat in the temperature interval considered.

2.7 High Tension

Without any power lines in the vicinity, the red-painted part of the compass needle normally points in the north direction. It will also do so close to power lines.

One might first think of the famous experiment in 1820 by the Danish physicist Hans Christian Ørsted, who found that a compass needle, when placed under a wire carrying a current, turns to a position perpendicular to the direction of the wire. However, the power lines over land (contrary to many underwater cables) almost always use alternating current. The frequency may be 60 Hz as in North America or 50 Hz as in most other parts of the world. This is much too fast for the compass needle to follow. The force from the power line averages out to zero, and only the Earth's magnetic field affects the compass. Moreover, the power lines usually have several cables in which there is a phase difference such that the combined magnetic effect on the compass needle is approximately zero at every instant.

What is called the north pole and the south pole of a compass needle (or any magnet) is determined by a convention, which may be related to the magnetic effects when a current passes through a wire. Soon after Ørsted's discovery, André-Marie Ampère formulated the "right-hand screw rule." It can be given several equivalent formulations. One of them reads: "With the thumb of the right hand pointing in the conventional direction of the current, the curled fingers around the wire indicate the circular sense of the magnetic field around the wire." Another formulation is: "Your right hand is held with your fingers along the wire, in the conventional direction of the current, and with the inside of the hand toward the magnet. The magnet's north pole deviates in the direction of your thumb" (see fig. 2.13). Maxwell's screw rule reads: "If a right-handed screw is turned so that it moves in the same direction as the current, its direction of rotation will give the direction of the magnetic field." Note that the convention for the north and the south pole of a magnet thus is related to another convention—that of the direction of a current, which is said to go from the positive to the negative pole of a battery. The actual charge transport in a wire by the negatively charged electrons is in the opposite direction.

There is a sport called orienteering where you are given a map

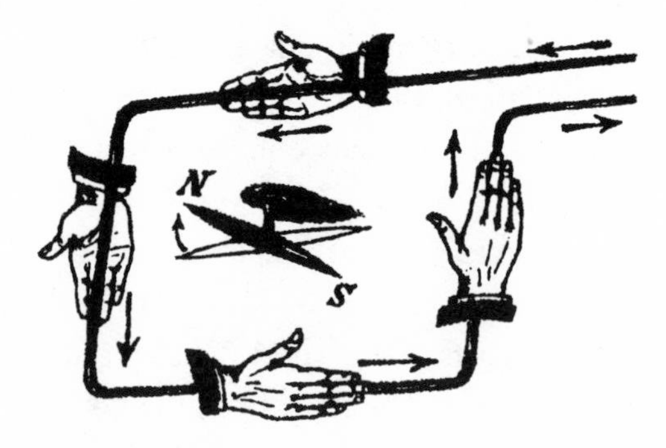

Fig. 2.13. One version of the right-hand rule

with marked checkpoints. The task is to complete the course defined by these checkpoints as fast as possible. The course often goes through wilderness; then you need a compass. But the same compass is not used in Europe and in Australia or New Zealand. The magnetic inclination is quite different in the Northern and the Southern Hemisphere. The needle should be balanced so that it takes a horizontal position and does not touch the top or bottom of the oil-filled capsule enclosing the needle when the compass is held flat.

Additional challenge. It is a convention to paint that part of the compass needle red that points toward the magnetic pole located in Canada. Is this red-painted end a magnetic north pole or a magnetic south pole? (See the solution at the end of this chapter.)

2.8 Ocean Surface

The ocean surface forms a small hump.

In a problem like this it is instructive to sketch a figure, with features exaggerated and without regard to accuracy in details. Figure 2.14 gives the two alternatives in the problem.

The gravitational force between two bodies only depends on their masses and not on what they are made of or whether they are solid or liquid. If the underwater mountain is replaced by water, the ocean surface would be flat (neglecting the curvature of the Earth). Because minerals have a higher density than water, the mountain pulls

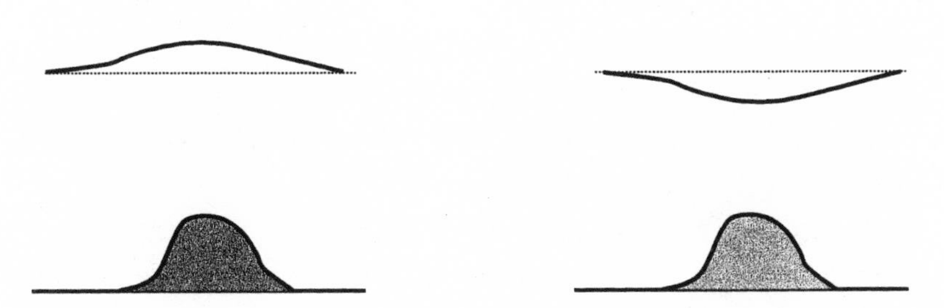

Fig. 2.14. Is the ocean surface raised or depressed over an underwater mountain?

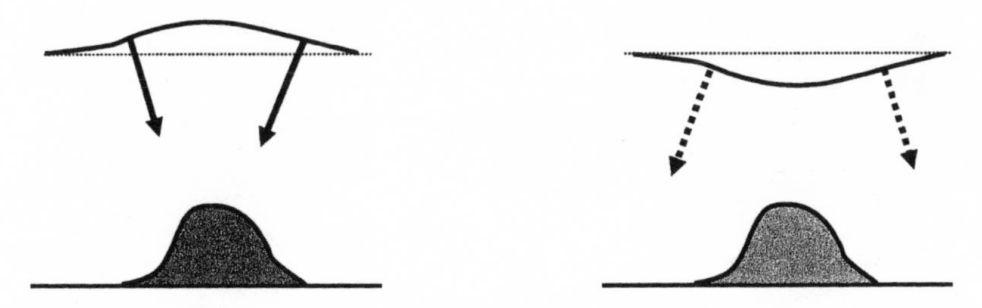

Fig. 2.15. The force acting on a small volume of water at the surface must be directed perpendicular to the surface.

with an increased gravitational force on the ocean water. That force must be perpendicular to the ocean surface, since any force component parallel to the surface would set the water in motion. It is now obvious, according to the schematic illustration (fig. 2.15) that the extra gravitational force from an underwater mountain gives rise to a small hump on the ocean surface.

How physicists think. The mathematically minded physicist would say that the shape $\phi(\boldsymbol{r})$ of the ocean surface represents an equipotential surface for the gravitational potential. The gravitational force on a water element near the surface is directed along the gradient $-\nabla\phi(\boldsymbol{r})$. Here $-\nabla\phi(\boldsymbol{r})$ is a vector, which forms a normal to the surface described by $\phi(\boldsymbol{r})$. In the analogous way, $-\nabla T(\boldsymbol{r})$ gives the direction of heat flow when $T(\boldsymbol{r})$ describes the temperature at points $\boldsymbol{r}$ in a body, and $-\nabla c(\boldsymbol{r})$ gives the direction of the diffusion flow caused by a gradient in the concentration $c(\boldsymbol{r})$. The minus sign in these expressions can be understood from the following special case. Consider the flow of heat between two points on the x axes in a Cartesian coordinate system. The flow is proportional to the temperature difference between the points, that is, it depends on the derivative $\mathrm{d}T/\mathrm{d}x$. But $\mathrm{d}T/\mathrm{d}x$ is positive if the temperature T increases along the x direction, while the heat flows in the opposite direction, from a high T to a low T.

The change in the ocean level above an underwater mountain (relative to what one would get for a smooth seabed) is much too small to be detected by the naked eye, but it can be measured by satellites. Such data are used in *bathymetry* to map out the topography of the seabed.

2.9 Mariotte's Bottle

The "constant flow rate" is larger if the tube T is raised somewhat.

The key to the solution is the fact that the pressure in a liquid is determined by the depth below a free surface where the liquid and the atmosphere are in direct contact. The upper part of the water column inside the tube T is in direct contact with the atmosphere, which gives a constant pressure there. When water flows out from the opening O, the pressure of the air trapped in the upper part of the bottle decreases because it is confined to a volume that becomes larger. Then the upper water level inside the tube is pressed down below that of the water in the bottle. Eventually the level inside the tube T reaches the lower end of the tube, and atmospheric air can enter the bottle. That is the start of the period with constant flow rate out from the opening O. The flow rate depends on the pressure difference between both sides of the opening. That pressure difference is determined by the vertical distance between the lower end of the tube T and the opening O.

Although we could solve the problem without the use of formulas, it is illuminating to add some mathematics to our argument. Let

p_{liq} be the pressure in the liquid just inside of the opening, p_{atm} is the ambient pressure, and ρ is the density of the liquid. The speed v of the liquid flow through the opening is given by

$$p_{\text{liq}} - p_{\text{atm}} = \tfrac{1}{2}\rho v^2$$

This is a special case of *Bernoulli's equation,* and essentially expresses the conversion of potential energy to kinetic energy. With distances as in figure 2.16, and when $x = h$,

$$p_{\text{liq}} - p_{\text{atm}} = \rho g h$$

and therefore

$$v = \sqrt{2gh}$$

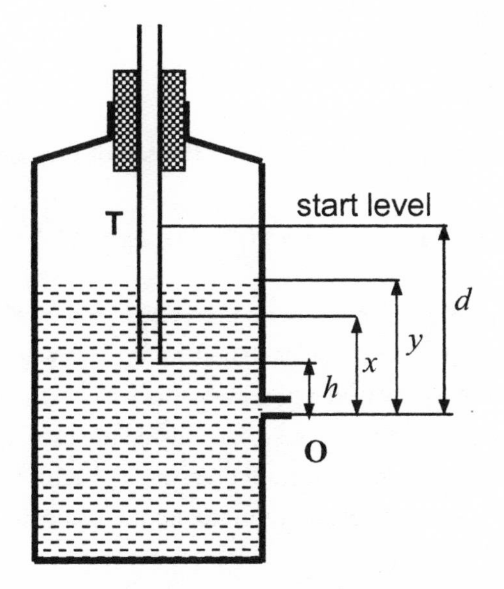

Fig. 2.16. Definition of lengths h, x, y, and d. The opening O is assumed to be small.

This result is known as *Torricelli's principle.* If the tube T is lifted up a little bit, the distance h increases, which gives a higher speed v and hence a larger flow rate. Our arguments assume that the opening O is in some sense "small." (We refrain from discussing what, more precisely, is meant by "small.")

The construction is known as Mariotte's bottle. Edme Mariotte (c. 1620–1684) was a French Roman Catholic priest and prior, who stated Boyle's law for gases independently of Boyle. In France it is therefore called Mariotte's law. The word barometer was coined by Mariotte. The constant flow rate from Mariotte's bottle was earlier of practical importance in laboratories. In the nineteenth century it was also frequently used in oil lamps for domestic illumination.

The various stages of the flow can be described in more detail as follows, and with distances defined as in figure 2.16. Recall the key principle that the pressure in the liquid at a certain point P is equal to the pressure p_0 at a liquid surface S in contact with the atmosphere plus a term $\rho g z$, where z is the vertical distance between P and S. The pressure at the bottom of the tube T is

$$p_0 + \rho g(x - h)$$

The pressure in the air trapped under the top of the bottle is

$$p = p_0 + \rho g(x - h) - \rho g(y - h) = p_0 - \rho g(y - x)$$

This is lower than the atmospheric pressure p_0, because the volume available to the trapped air inside the bottle increases when water leaks out through O. We now understand why the water surface inside the bottle lies higher than the water surface inside the tube T, which is in direct contact with the ambient air of pressure p_0. In this situation, also shown in figure 2.16, the pressure in the liquid just inside O is

$$p_0 + \rho g x$$

The flow rate is proportional to the pressure difference between the two sides of the hole, that is:

$$(p_0 + \rho g x) - p_0 = \rho g x$$

As water leaves through O, x decreases and so does the flow rate. When x has reached the lower end of the tube, air from the ambient atmosphere can bubble up into the air trapped at the top of the bottle. The pressure at the top then becomes

$$p_0 - \rho g(y - h)$$

This pressure increases as water continues to flow out from the bottle and the depth y becomes smaller. However, the pressure difference between the two sides of the opening O,

$$(p_0 + \rho g h) - p_0 = \rho g h$$

is constant. Therefore also the flow rate from O now becomes constant. It depends on how far down the tube T reaches, as given by the distance h. This situation prevails until the liquid surface inside the bottle has reached the lower end of the tube. When that happens, the air inside the bottle is in direct contact with the ambient atmosphere, and thus has the constant pressure p_0. The flow rate decreases gradually until the flow stops when the level inside the bottle has reached the opening O.

ADDITIONAL CHALLENGES

2.3 Bright Lamps?

The resistor should be inserted between the left side of the lamp and the right side of the battery. Then we have a balanced "bridge" with no potential difference over the lamp.

2.4 Low Pressure

The mother is heavier and therefore compresses the tire more. This gives a smaller effective wheel radius and a longer recorded distance. But if the measuring device is mounted on the front wheel, and either the daughter or the mother has difficulties keeping the front wheel on a straight path, that may be a dominating cause for different measured distances.

2.7 High Tension

The red-painted end is a magnetic north pole, pointing toward the *geographic* north pole, which is a *magnetic* south pole.

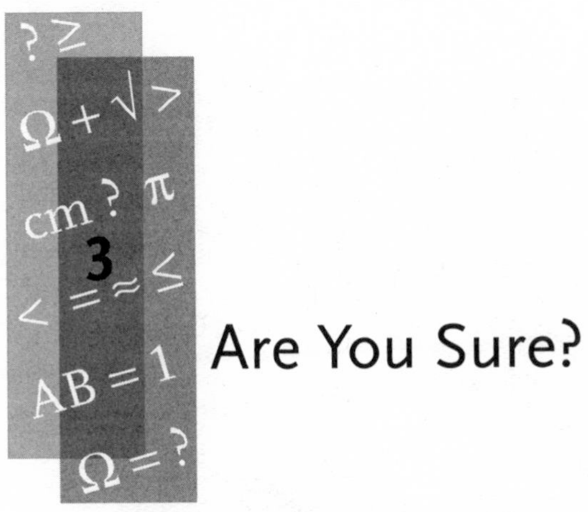

Are You Sure?

The problems in this chapter may seem simple. But beware. For some readers the answer could be a real surprise. The last problem may baffle even a physics professor.

PROBLEMS

3.1 Bicycle on a Rope

This problem has been used as a part of team-building activities, because everyone can have an opinion, and the opinions often disagree. A string is attached to a pedal of an ordinary ungeared bicycle, when the pedal is in its lowest position. You stand behind the bicycle (let someone hold it upright) and pull the string horizontally. Does the bicycle start to move forward or backward, or will it move with the tire of the rear wheel sliding on the road? How is the pedal moving relative to the bicycle frame? Does it rotate forward (like in an ordinary bicycle ride) or does it rotate backward?

3.2 Boats in a Lock

How much less water is required to pass a *large* boat through a lock than a *small* boat? To be more precise, consider two lakes at different elevations, connected by a lock (fig. 3.1). A boat enters through the open lock gate in the lower lake, and the gate is closed. The lock

Fig. 3.1. Old-fashioned style lock

chamber is filled with water from the upper lake, until the level in the lock is the same as in the upper lake. Then the upper lock gate is opened, and the boat can continue its voyage.

Compare two such passages of a boat through the lock. In one case the total mass of the boat is 50 tons, and in another case the mass is only 5 tons. How much more water, expressed in cubic meters, has to flow into the lock from the upper lake when the small boat passes through the lock than when the large boat passes through?

3.3 Humming Transformer

A transformer gives a humming sound. Here is a question for you who live in North America. What is the frequency of the humming in London? All others should tell what the frequency is in New York.

3.4 What Is the Charge?

The SI unit for the capacitance of a capacitor is farad, written F. One farad is a *very* large capacitance, and it is not unusual to find capacitors rated 1 pF. The prefix pico (p) represents 10^{-12}. Similarly, the prefix nano (n) represents 10^{-9}. What is the charge Q, in coulomb

(C), of a capacitor with the capacitance $C = 1$ pF when it is charged to the voltage $U = 1$ nV?

3.5 Two Wooden Blocks

A block of wood (a parallelepiped) with dimensions 5 cm × 10 cm × 30 cm floats in water. The vertical distance from the water surface to the highest point of wood is 2.5 cm. Then an identical block is placed on top of the first block (fig. 3.2). What is now the vertical distance between the water surface and the top of the second block?

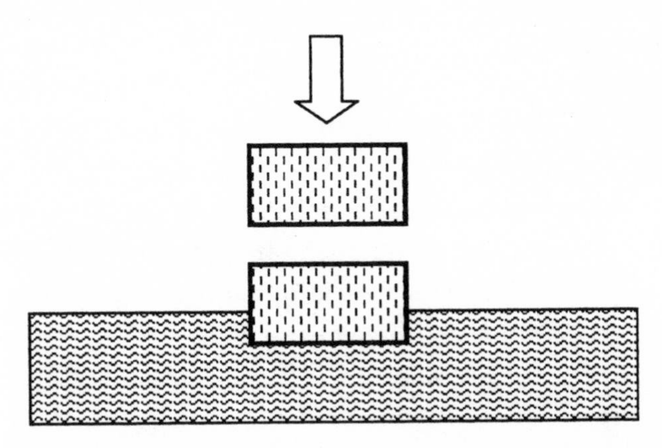

Fig. 3.2. A block of wood floats in water, and an identical block is put on top of it. How high above the water surface is the top of the second block?

3.6 Shot in a Pot

A cylindrical 6-liter aluminum pot floats in a bathtub filled with water. The mass of the pot is 1 kg. Place a woman's shot from shot putting in the pot (fig. 3.3). According to the athletics rules, the mass of the shot is 4 kg. How high above the water surface in the bathtub is the rim of the pot?

3.7 Filling a Barrel

An open water barrel can be filled from the top through a faucet, and emptied through a hole at its bottom. With the faucet closed, a full barrel is emptied in 9 min. With the hole closed, an empty barrel is

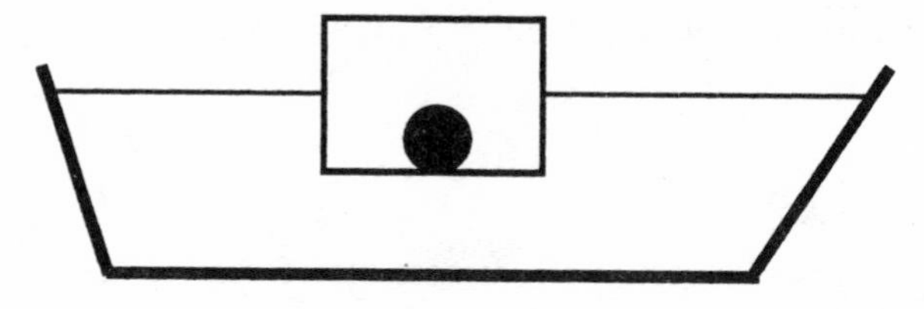

Fig. 3.3. Schematic illustration of a shot in a pot that is floating in a bathtub

filled in 8 min. How long does it take to fill an empty barrel, if both the hole and the faucet are open?

3.8 Tube with Sand

A very long, vertical tube with a radius of 5 cm (2 in) can be covered with a light disk at its lower end (fig. 3.4). You stand below the tube and press the disk onto the empty tube. Then the tube is filled with dry sand through its upper end. Are you strong enough to prevent the sand from flowing out, when the pile of sand in the tube is 10 m (33 ft) high? The tube itself is firmly anchored to a support.

Fig. 3.4. A tube, with its lower end covered by a disk, is filled with sand.

3.9 Sauna Energy

A sauna is like a room where the temperature of the air is very high. In one case a small private sauna is heated with an electric unit rated at 6000 W. After 20 min the unit is automatically shut off because the temperature has reached a preset value of 85 °C (185 °F). What is the increase in the energy of the air inside the sauna?

3.10 Slapstick

In *Slapstick,* a novel by the American writer Kurt Vonnegut, the world experiences sudden jolts of high gravity forces. In Chapter 31, we are told how New York is hit.

> The force of gravity had increased tremendously. There was a great crash in the church. The steeple had dropped its bell. Then it went right through the porch, and was slammed to the earth beneath it.
>
> In other parts of the world, of course, elevator cables were snapping, airplanes were crashing, ships were sinking, motor vehicles were breaking their axles, bridges were collapsing, and so on and on.
>
> —*Slapstick, or Lonesome no More!* (New York: Bantam Doubleday Dell, 1976 [1989])

This passage has caught the interest of physicists. Assuming that gravity changed as described, is there anything in Vonnegut's text that stands out as unphysical?

SOLUTIONS

3.1 Bicycle on a Rope

The bicycle moves backward on the road, while the lower pedal moves "forward" relative to the bicycle frame, that is, as it would move if you try to pedal the bicycle backward.

Rather than dealing immediately with the situation in the problem, we first consider the normal use of a bicycle. When one pedal is in its lowest position, that pedal moves *backward* relative to the *bicycle frame,* while the bicycle moves forward. But for any normal gear, the circular path followed by the pedal is shorter than the distance on the road covered by the bicycle itself. Therefore the pedal moves *forward* relative to the *road*. This is the result we have already encountered in problem 2.1. In short, when the bicycle moves forward on the road, a pedal in its lowest position moves backward relative to the frame. Conversely, when the bicycle moves backward on the road, a pedal in its lowest position moves forward relative to the frame. This gives the answer to our problem, because if you pull the rope toward you (i.e., shorten the distance between you and the bike), it is obvious that the bicycle frame moves backward on the road, with the rear wheel either rotating or sliding on the ground.

In normal bicycle gearing the wheel would not get stuck and slide at the start of this motion.

3.2 Boats in a Lock

Intuitively one might think that more water is needed in the case of a small boat, but in fact the amount of water is the same, irrespective of the size of the boat.

A simple argument is as follows. The boat has entered the lock chamber, and the lower lock gate is still open. Then the water levels in the chamber and in the lower lake are the same. We now close the lower gate, so that both gates are closed. Suppose that the difference in elevation between the two lakes is 2 m. We can raise the level of the water in the lock chamber to that of the upper lake by inserting a "slice" of water, 2 m high, at the bottom of the lock. As a rule, if the horizontal area of the lock chamber is A and the elevation difference is h, we must fill the lock with the volume Ah of water each time boats are passing through the lock, irrespective of how many there are, or how big.

3.3 Humming Transformer

In London the main humming frequency is 100 Hz, and in New York it is 120 Hz.

The frequency of the alternating current (AC) and voltage in the United States is 60 Hz. In most other parts of the world, including Europe, it is 50 Hz. During each cycle, the voltage reaches a positive and a negative peak value. For 50 Hz AC, the voltage thus has a maximum value 100 times each second. The origin of the humming is a phenomenon called magnetostriction. A transformer has a core with many sheets of steel with special magnetic properties. During each voltage cycle the sheets are extended and contracted twice, because the magnetostriction depends only on the magnitude of the instantaneous voltage, and not on its "sign."

Any caveats? No vibrating system is perfect. Therefore the transformer also gives rise to humming at frequencies that are odd multiples of the fundamental frequency. In London this means 3×100

$= 300$ Hz, $5 \times 100 = 500$ Hz, and so on; and in New York this means $3 \times 120 = 360$ Hz, $5 \times 120 = 600$ Hz, and so on. The importance of the higher frequencies in this series rapidly decreases, and it is reasonable to say that the humming frequency is 100 Hz (120 Hz).

One may ask if there is another explanation for the humming vibrations that is not related to magnetostriction. Between two parallel wires, each carrying a current I, there is a force F given by (compare *Biot-Savart's formula*)

$$F = \frac{L\mu_0 I^2}{2\pi d}$$

Here $\mu_0 = 4\pi \times 10^{-7}$ Vs/(A·m) is the permeability of a vacuum. L is the length of each of the wires, which are separated by the distance d. Perhaps such a force, acting between the wires in the transformer coils, also excites vibrations in the transformer? If that is so, the effect should increase with the load on the transformer, that is, with the current in the coils. This reasoning is correct but it turns out to be a small effect compared with that from magnetostriction. Usually, the difference between no load and full load is a change in the sound level of less than 1 to 2 dB.

3.4 What Is the Charge?

This is impossible with an ordinary capacitor, because the charge calculated by the standard relation, $Q = CU = 10^{-21}$ C, is smaller than the charge $e = 1.6 \times 10^{-19}$ C of a single electron.

To illustrate that 1 pF is not an unreasonably small value, we can consider the capacitance of a sphere. It scales as its radius R. The capacitance 1 pF corresponds to approximately $R = 1$ cm. The capacitance of a sphere having the size of the Earth is about 600 μF.

In a famous experiment Robert A. Millikan found that the electric charge of small oil droplets was an integral number of an elementary charge $e = 1.6 \times 10^{-19}$ C. He interpreted e as the charge of an electron. For this achievement, Millikan was awarded the 1923 Nobel Prize in physics.

In elementary particle physics we learn that the proton and the neutron are composed of quarks, which have fractional charges, $2e/3$ or $-e/3$. However, the quarks are bound together in such a way that free charges are always integer multiples of e. In certain devices in microelectronics, one speaks of noninteger charges, but this is a subtler concept connected with small spatial displacements of charge and does not refer to the commercial type of capacitor in our problem.

Outlooks. In our relation $Q = CU$ for the charge Q of a capacitor charged to the voltage U, we used the symbol C for the capacitance. The resulting charge Q was expressed in the SI unit coulomb, denoted C. In handwriting it would be difficult to distinguish between C meaning capacitance and C meaning the unit coulomb, but not so in printed text. Symbols printed in italics denote a *quantity*, which means a combination of a *numerical value* and a *unit*, for instance $C = 25$ pF. Units are always written in an upright font. The equation editor in some computers automatically writes symbols in italics, except for letter combinations like sin or log, which denote mathematical functions and therefore are written with an upright font.

Our problem also dealt with prefixes. Those that refer to negative powers of 10 are as follows.

Name	Symbol	Power of 10
deci	d	10^{-1}
centi	c	10^{-2}
milli	m	10^{-3}
micro	µ	10^{-6}
nano	n	10^{-9}
pico	p	10^{-12}
femto	f	10^{-15}
atto	a	10^{-18}
zepto	z	10^{-21}
octo	o	10^{-24}

The prefixes deci, centi, and milli come from the Latin for ten, hundred, and thousand. Micro is Greek for small. Nano is often erroneously thought to stand for nine, but its Greek and Latin roots mean dwarf. Pico comes from Greek for very small. Femto and atto come from Scandinavian words for fifteen (femten) and eighteen (atten). Zepto and octo are derived from Greek words for seven and eight.

3.5 Two Wooden Blocks

Very likely, the distance is still 2.5 cm. The orientation with a horizontal upper surface is unstable for a bar with a square cross section that floats half submerged in water. The combined block rotates 45° and the upper block slides off, so that the two blocks are floating side by side.

Consider a long parallelepiped, with a square cross section and the density ratio

$$x = \frac{\rho_{\text{wood}}}{\rho_{\text{water}}}$$

It is a straightforward, but rather lengthy, task to find the orientation that minimizes the total energy when such a bar floats in water. Let the angle between a side of the square bar and the water surface be ϕ. When the density ratio x is close to 0 or 1, the bar floats with one of its sides in a horizontal orientation, that is, $\phi = 0$. The variation in ϕ with x is shown in figure 3.5. The angle ϕ is zero from $x = 0$ up to a value $x = x_1$. At $x = x_1$, $\mathrm{d}\phi/\mathrm{d}x$ is infinite. The orientation of the bar thereafter changes gradually in the interval $x = x_1$ to $x = x_2$. A detailed analysis shows that $\mathrm{d}\phi/\mathrm{d}x$ is infinite at the following values of the density ratio x:

$$x_4, x_1 = \frac{3 \pm \sqrt{3}}{6} \approx 0.5 \pm 0.29$$

$$x_3, x_2 = \frac{16 \pm 7}{32} \approx 0.5 \pm 0.22$$

The angle ϕ is zero in the intervals $0 < x < x_1$ and $x_4 < x < 1$. It is 45° in the interval $x_2 < x < x_3$,

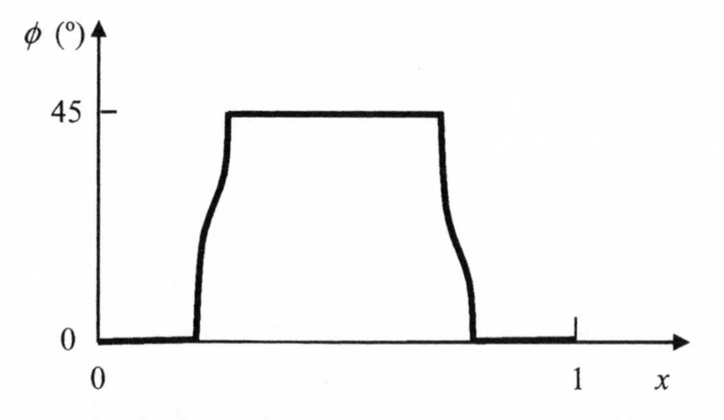

Fig. 3.5. Angle ϕ for the equilibrium orientation, as a function of the density ratio $x = \rho_{solid}/\rho_{liquid}$

How physicists think. Let us start with $x \approx 0$. At this value, the bar floats with a horizontal upper surface. Between $x = 0$ and $x = x_1$, it remains horizontal ($\phi = 0$ and $d\phi/dx = 0$). At $x = x_1$ there is a sudden change to an infinite $d\phi/dx$. Most physicists, relying on experience from other examples of instabilities, would probably take the infinite value of $d\phi/dx$ as indicating that ϕ abruptly changes to 45°. But not so! There is a *gradual* change in the angle ϕ as x varies from x_1 to x_2. In this case even an experienced physicist may be misled by her intuition.

Any caveat? If you try this experiment several times, it is possible that you get the result in our solution only the first time. When the wooden surfaces are wet, they tend to stick together. With the two blocks placed on top of one another, they turn around in the water but perhaps they only partly separate.

Outlooks. Icebergs can be very unstable. In *Twenty Thousand Leagues Under the Sea* (1870) Jules Verne wrote how the submarine Nautilus was hit. Captain Nemo explained: "An enormous block of ice; a mountain turned over. When icebergs are undermined by warmer waters or by repeated collisions, their center of gravity rises, with the result that they overturn completely."

Everyone knows that the "the tip of the iceberg" contains only a small part of the iceberg's total volume. The density of freshwater ice at 0 °C is 917 kg/m^3. Real ice often has a lower density because of trapped air bubbles. The salinity and temperature of ocean water at the surface, and therefore also the density, vary with the geographic location. In cold ocean water, where you may find icebergs, the density typically is about 1025 kg/m^3. Then, with an assumed density 900 kg/m^3 for the ice in icebergs, 1/8 of the iceberg's mass is above the surface of the ocean. But when this is illustrated with a floating iceberg, the artist sometimes makes several errors. If an approximately conical iceberg floats as on the left side of figure 3.6, with 1/10 of its *height* above the surface, only 1/1000 of the volume is above the surface. On the right side of the figure, with 10 % of the shown *area* above the surface, 3 % of the *volume* is above the surface. Moreover, a conical iceberg will not float with its axis of symmetry in the vertical direction. You could try an experiment by using a funnel as a mold to make a conical iceberg in your freezer, and then check how it floats.

Archimedes (c. 287–212/211 BC) wrote two books entitled *On Floating Bodies*. The first book contains a proof that the segment (top

Fig. 3.6. These two schematic drawings of floating icebergs contain several errors.

slice) of a homogeneous solid sphere always floats with its base parallel to the surface, either above or below it. The second book contains a detailed account of the equilibrium positions of floating bodies shaped as paraboloids, as a function of the density ratio of the body and the fluid. These are truly remarkable achievements, considering that Archimedes did not have the mathematical tools that were not developed until about 2000 years later. Even our problem with the stability of a bar with square cross section leads to such long calculations that the steps in the mathematical solution are not given in this book.

3.6 Shot in a Pot

The pot sinks to the bottom of the bathtub.

Many people would, incorrectly, argue like this. The total mass of the pot and the shot is 5 kg. According to Archimedes' principle the floating pot must displace 5 kg of water, that is, the volume 5 L. (The density of water is 1000 kg/m^3 or 1 kg/L, where L, or l, is the SI symbol for liter.) Since the volume of the pot is 6 L, we could load it with an additional 1 kg before it sinks. But we seem to get stuck here, because we don't know the bottom area (or the height) of the pot—only its volume.

Usually a figure is helpful when we try to solve a problem in physics, but in this case it could be misleading. A spherical shot placed at the center of the pot gives a dynamically unstable system. With the slightest deviation to one side or another, the sphere tilts the pot and rolls to its side. The edge of the pot comes below the water surface, and the pot and the shot sink to the bottom of the bathtub.

Any caveat? Now you should be critical and ask if this always happens. Apply the problem-solving technique of going to an extreme limit. Imagine a 6-liter pot that has the same diameter (about 10 cm) as the shot. Its height is 76 cm! Such a pot floats with vertical walls and is submerged to 5/6 in the water, when it contains the shot. On the other hand, it is so tubelike that it falls to the side and sinks if placed unloaded in the bathtub—or does it? Could water flow into a "tube" so that it rises? (You might like to try this with test tubes.) Such a shape is not what we would call a pot, however. It is left as a challenge to the reader to analyze if there is a cylinder with height-to-diameter ratio such that it can float unloaded, as well as loaded with the shot. (The book gives no answer to this challenge, but see "Further Reading.")

3.7 Filling a Barrel

The barrel will never be full.

We first give an erroneous "solution" of a type that can be found in some books with mathematical puzzles, and even in schoolbooks in mathematics. In 72 min, the inflow of water from the faucet suffices to fill $72/8 = 9$ barrels, but the outflow during the same time corresponds to only $72/9 = 8$ barrels. The difference is exactly one full barrel. Thus it takes 72 min to fill the empty barrel. But this answer is wrong. The barrel will never be filled in our numerical example!

The error lies in the assumption that the flow of water through the hole is independent of the depth of the water. In fact, the outflow (volume per time) varies as the square root of the water depth. The initial rate of flow from a *full* barrel would empty it in much less than 9 min, if that rate stayed unchanged. Thus, the outflow from a *nearly* full barrel is (in our example) larger than the (constant) inflow. On the other hand, the flow rate from the barrel is very low when the barrel is nearly empty.

Let us now start with an empty barrel, and with both the hole and faucet open. The water level first rises. When it has reached a certain height, the level remains constant because the rates of inflow and outflow of water are equal. A simple mathematical model (but beyond the scope of this book), which assumes that the outflow varies as the square root of the depth of water, shows that with the data in our example the barrel will be filled to $81/256$ of its height.

Of course we can imagine other data for the inflow and outflow such that the barrel will eventually spill over. In all cases a more detailed calculation is necessary, which also takes into account the geometry of the barrel. Whether the barrel is narrow and high, or wide and low, makes no difference in the time to *fill* it, but affects the time to empty it.

The rate of flow from a cylindrical vessel with a hole in the bottom can be modeled with *Bernoulli's equation* for fluids. If the initial height of the liquid in the vessel is H, the acceleration of gravity is g, the area of the free liquid surface in the vessel is A and the (effective) area of the hole is a, one can show that the time t to empty the vessel is

$$t = \left(\frac{A}{a}\right)\sqrt{\frac{2H}{g}}$$

Note that the density of the liquid does not enter into this calculation. It takes the same time for water as it takes for any other liquid, provided that the viscosity of the liquid is not important. For a barrel with height $H = 1$ m we have $(2H/g)^{1/2} = 0.45$ s. In our example it took 9 min = 540 s to empty the barrel. If its height is 1 m, we get $a/A \approx 1/1200$, which is a realistic value. For instance, if the barrel is 0.9 m (3 ft) wide, the effective diameter of the circular drainage hole is 26 mm (about 1 in).

3.8 Tube with Sand

There should be no problem in pressing on the disk hard enough to prevent the sand from coming out, when the tube is as narrow as in our example. The force needed is less than that from only a few kilograms of sand.

Your first thought might have been that one must press against the entire weight of the filling in the tube, as one must do if the sand is replaced by water (compare fig. 3.7). With a tube radius $R = 0.05$ m, height $H = 10$ m, typical density of the minerals in sand $\rho = 3000$

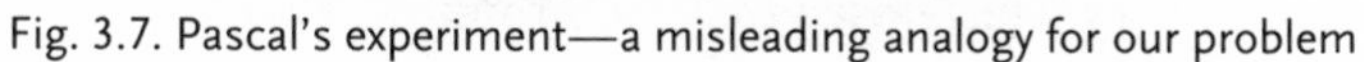

Fig. 3.7. Pascal's experiment—a misleading analogy for our problem

kg/m^3 and reasonable packing fraction $\eta = 0.6$, the mass in the filled tube is

$$\pi R^2 H \rho \eta \approx 140 \text{ kg}$$

This is more than most people can hold back.

But our argument neglects the friction between the sand grains and the inner tube wall. Again we could compare this with a tube filled with water. The water pressure at a certain depth is isotropic, that is, there is a component also normal to the tube wall. A similar mechanism presses the sand grains against the tube wall, and allows a vertical friction force to come into play. A theoretical model (which is not further discussed here) gives the vertical stress (pressure) on the disk as

$$\sigma = \rho \eta g \lambda (1 - e^{-H/\lambda})$$

Here g is the acceleration of gravity. The expression also contains a *characteristic length parameter,* λ. It is defined as

$$\lambda = \frac{kR}{f}$$

where k is a dimensionless constant of the order of 1 and f is the friction factor between sand grains and the tube wall. When $H >> R$ (as in our example) we can ignore the exponential term in the expression for σ. Then we must press on the disk (which has the area πR^2) with the force $\sigma(\pi R^2)$ to prevent the sand from coming out. In other words, we must hold back a load, which is the weight of the mass

$$M = \frac{\sigma \pi R^2}{g} = \pi R^2 \rho \eta \lambda \sim (\pi R^2 \rho \eta) R$$

In the last step we have used that, very roughly, $\lambda = kR/f \sim R$. (The symbol ~ here means "of the order of.") Thus, we don't need to hold back a pile of sand with height H but rather a pile with height of the order of the radius of the tube R. The corresponding mass is a factor of the order of $R/H = 0.005$ smaller than the mass of 140 kg that we first obtained, or about 1 kg (2 pounds). A small child could hold back the disk, even if the characteristic length λ turns out to be several times larger than R.

How physicists think. In terms of the mass, water is the most common substance handled by man. But after water comes granular materials, ranging from sand to grain. Despite this, the mechanical properties of granular materials are usually ignored in mechanics courses. However, as we shall now further illustrate, the concepts taught in the elementary mechanics courses are sufficient to understand the essential ideas of our problem. Its theoretical foundation, which is due to the German engineer H. A. Janssen and dates back to 1895, is important, for example, in the design of silos.

The stress σ was assumed to be given by the relation

$$\sigma = \rho \eta g \lambda (1 - e^{-H/\lambda})$$

which was quoted without derivation. In the scientific field of granular materials it is a standard result, but one should never just accept a formula, even if it is taken from a reliable source. There is always the possibility of misprints.

As a first check we verify that the relation is dimensionally correct. The factor $\rho \eta g \lambda$ has the dimension of a stress (pressure), or force per area, as is easily checked if we represent each quantity in the product $\rho \eta g \lambda$ with its SI unit, that is,

$$[\mathrm{kg/m^3}]\,[1]\,[\mathrm{m/s^2}]\,[\mathrm{m}] = [\mathrm{kg{\cdot}m/s^2}]\,[1/\mathrm{m^2}] = [\mathrm{N/m^2}]$$

Here [1] represents the dimensionless factor η. The identity 1 N = 1 $\mathrm{kg{\cdot}m/s^2}$ perhaps is best understood with reference to Newton's law $F = ma$. In our dimensional consideration of Janssen's expression for σ, we also notice that the argument H/λ in the exponential function is dimensionless, as it must be.

A second check is to consider limiting cases. Let us take a very small height H, so that the exponential term is well approximated by the first two terms in a series expansion; $e^{-H/\lambda} \approx 1 - H/\lambda$. Then

$$\sigma = \rho \eta g H$$

In this limit we must hold back sand with a force that corresponds to the weight of the mass

$$M = (\pi R^2 H)\rho\eta$$

that is, the actual mass of the short sand pile. This is exactly the result we expect when the height H of the pile is very small, and therefore the friction against the inner tube wall is not important.

Another limiting case is to let the friction factor f go to zero. Then the length parameter λ becomes very large. However, we may also want the height H to be large, so we make the additional assumption that $\lambda >> H$. This would correspond to the situation that we have a *given* tube with length H, where H could be large, and *then* take the limit of a *very* small friction factor f so that $\lambda >> H$. We get

$$1 - e^{-H/\lambda} \approx 1 - (1 - H/\lambda) = H/\lambda$$

The expression for σ takes the form

$$\sigma = \rho\eta g H$$

It is the same as we obtained above for a small H, but in this case H is large and the friction factor f is small. The required force on the disk is

$$\sigma\pi R^2 = (\pi R^2 H)\rho\eta g = Mg$$

It corresponds to the entire weight of the sand pile—an expected result when the friction between the sand grains and the tube wall is negligible. Note that we cannot directly let $\lambda = \infty$ when $f \to 0$, but must make a series expansion of the term $e^{-H/\lambda}$ before we take the limit.

Checking dimensions and checking limiting cases is the natural habit of an experienced physicist or engineer, but is sadly ignored by many young students.

3.9 Sauna Energy

The surprising answer is that the total energy of the air inside the sauna is unchanged and equal to the energy at room temperature.

We use two relations for classical gases. One is the equation of state that connects the volume V, the pressure p, the number of molecules N, and the thermodynamic (absolute) temperature T,

$$pV = Nk_B T$$

Here k_B is Boltzmann's constant. (Another equivalent relation is obtained with $R = N_A k_B$, where R is the gas constant and N_A is Avogadro's constant.) Furthermore, a molecule of a diatomic classical gas has the energy $3k_B T/2$ from the translational motion, that is, the energy associated with the kinetic energy of the molecule considered as a single particle. In addition there is an energy associated with the rotation of the molecule, like a freely rotating dumbbell. That gives an additional $k_B T$. The total energy per molecule is

$$E = \frac{5}{2} k_B T$$

The total energy of N molecules in the sauna is $E_{sauna} = NE$. We now get

$$pV = \frac{2}{5} E_{sauna}$$

The volume V of the sauna is certainly not changed as the sauna is heated. Also the pressure is unchanged, and equal to the atmospheric pressure outside the sauna. Therefore the energy E_{sauna} does not depend on the temperature T. So where does the energy go? Part of it is used to heat the walls and objects inside the sauna. Another part goes to the outside air. As the temperature increases inside the sauna, the average energy of an individual molecule increases. But at the same rate molecules leave the sauna, to keep the pressure constant. These two effects exactly compensate each other.

The traditional Finnish sauna is heated by a wood fire unit, and with a heap of stones that get very hot. Nowadays electric heating is much more common. Water is poured onto the stones to increase the humidity in the air. In that case the temperature is usually kept somewhat lower than in a dry-air sauna. Taking a sauna bath is common in countries like Finland and Sweden. Many one-family houses have their own sauna. It is comfortable to stay in the temperature of about 85 °C for 10–15 min, and this is the temperature we will use in the discussion below. Contrary to what is sometimes thought by those who have never experienced a sauna, such a high temperature is no health hazard. Some brave people like to plunge themselves into a cold lake after the sauna, perhaps in a hole in the ice. A cold shower can be a substitute.

Any caveat? No room is so tight that the pressure increases when the air temperature increases, as is obvious from the following example modeled for a sauna. Assume that the temperature increases by 65 °C, from 295 K to 360 K for air in a tight room. The pressure changes by a factor of 360/295, that is, an increase by about $(65/295) \times 10^5$ Pa $\approx$ 22 000 Pa if we start with the normal atmospheric pressure 10^5 Pa. That gives an increased force on the walls that is 22 000 N/m^2. The area of a sauna door typically is a little more than 1 m^2. Thus, the air in an assumed completely tight sauna would push on the door with the extra force corresponding to the weight of about a big car! A related comment can be made regarding the pressure inside a living room, as the temperature rises when an intense sunshine makes it warmer indoors. A change by 5 °C could increase the force on a big windowpane by about 3 kN, or the weight of more than three people. It is obvious that houses cannot be so tight that there is a pressure difference between the outside and the inside air of this magnitude. However, wind gusts can give rise to significant, but lower, pressure differences. Glass in windowpanes is surprisingly elastic and bends under such a pressure difference.

Any caveat? Let us return to the energy considerations. In schoolbook discussions one usually focuses on *monatomic* ideal gases. Under constant volume such a gas has an internal energy U that can be written

$$pV = \frac{2}{3}U$$

We used the prefactor 2/5 instead of 2/3 above because we have diatomic molecules. One may wonder if we should not have used a prefactor 2/7. The heat capacity per molecule and at constant volume is $c_V = 5k_B/2$ and the heat capacity at constant pressure is $c_p = 7k_B/2$. In our case the pressure was constant, but the quoted relation for c_p is derived for the condition that a fixed amount of gas is contained in a vessel that can expand and do work on the surroundings. That is not the condition in our problem.

The extra energy $2 \times (k_B T/2)$ for a diatomic molecule, which comes from the two additional degrees of freedom describing rotations of the molecule, is the high temperature approximation in classical physics to the more correct quantum mechanical description. It turns out to be a very good approximation to treat the diatomic molecules N_2 and O_2 as we have done, in the temperature range of interest for the sauna problem. At much higher temperatures there would also be a contribution from the vibrations of the atoms relative to the center of mass of the molecule. Furthermore there is a small amount of triatomic molecules in air, mainly CO_2 and H_2O. At the temperatures of interest here, the energy of CO_2 is better approximated by $7k_B T/2$. CO_2 makes up the same fraction of the total number of particles in the air inside and outside the sauna so our argument about the energy holds also in that case. But to be more precise, there *is* a correction to the answer of our problem because the concentration of H_2O is likely to be higher inside the sauna, and the average energy of a H_2O molecule is different from that of the dominating diatomic molecules. However, this effect is small in our context.

The sauna temperature of 85 °C, which was used above, refers to that level where one is sitting or reclining on a bench. Close to the floor it could be cooler by 20 °C or more, depending on how the ventilation is arranged. That does not change our conclusion because the energy per volume,

$$\frac{U}{V} = \frac{5}{2}p$$

is the same in all parts of the sauna, irrespective of variations in the temperature.

One final comment. In physics, energy does not have a unique absolute value. We can only define *energy differences.* (Unlike, for instance, the entropy S, which has a definite value because the third law of thermodynamics says that $S = 0$ at $T = 0$ K.) Not even the kinetic energy has a definite value unless we specify the reference frame. A seated passenger in a moving train has zero kinetic energy relative to the train but not relative to the ground. Tables of thermodynamic data for substances may give values of the enthalpy H as a function of temperature, for example. The numerical value of H requires a convention for the zero point, for instance, that elements in their most stable form at 25 °C have $H = 0$. When we discussed the energy per molecule it was assumed that the kinetic energy should be counted as zero for zero speed relative to the sauna, and the rotational energy should be considered as zero for non-rotating molecules. With these natural definitions, our result is valid.

3.10 Slapstick

Archimedes' principle says that the buoyancy force is equal to the weight of the displaced liquid. Therefore, as the weight of the ship increases in proportion to the force of gravity, so does the buoyancy, and the ship would not sink deeper. This is the physics schoolbook result. But in the real world of engineering, a large ship will probably sink.

Let us start with the observation that elevator cables break. Such a cable is designed to carry the elevator with its full load, and with a "factor of safety" that can be defined as the ratio

$$\frac{\text{actual strength}}{\text{required strength for allowed load}}$$

The required factor of safety in building codes for elevators is not a universal number, but varies with, for example, the design speed of the lift. Typically it lies between 7 and 12 (compare table 3.1). If cables break as Vonnegut writes, the increase in gravity force must be at least a factor of 10. This is a reasonable lower limit, because few elevators would happen to be loaded to their allowed limit when the jolt of increased gravity comes, and Vonnegut's text suggests that breaking cables was not uncommon. The increased load on the elevator cables during acceleration and retardation is not important in this context. The corresponding acceleration is certainly much smaller than g and therefore negligible compared with the change in gravity force that Vonnegut mentions.

Next we consider the forces acting on a large ship under normal gravity conditions. The buoyancy comes from the pressure of the water. That force is evenly distributed over the ship's hull. The weight of the ship, on the other hand, is the sum of the gravity forces acting on all the masses making up the ship. They are very unevenly distributed. In particular, just like church bells fall down and axles

Table 3.1. Minimum factors of safety in the California Code of Regulation for passenger and freight elevators

Rope speed		Safety factor	
ft/min	m/s	Passenger	Freight
200	1.0	8.40	7.45
600	3.0	10.70	9.50
1000	5.0	11.55	10.30
1500	7.6	11.90	10.55

break on land, we expect similar breaking and distortion in a large ship. The damages to the ship are likely to be severe enough that the ship sinks.

As a contrast we can think of a dinghy or a toy ship. It is so small that it withstands a gravity increase without internal breaking or distortion, and it floats at the same level as it did before the gravity change.

How physicists think. Understanding the schoolbook version of Archimedes' principle is important. But as we have seen in this example, engineering reality may have surprises for the unaware physicist, who has only focused on general principles. In particular, a scaling in size (up or down) may take you into a regime of new and important physical phenomena. For instance, Archimedes' principle is of little importance for an insect on a water surface, where surface tension dominates over buoyancy.

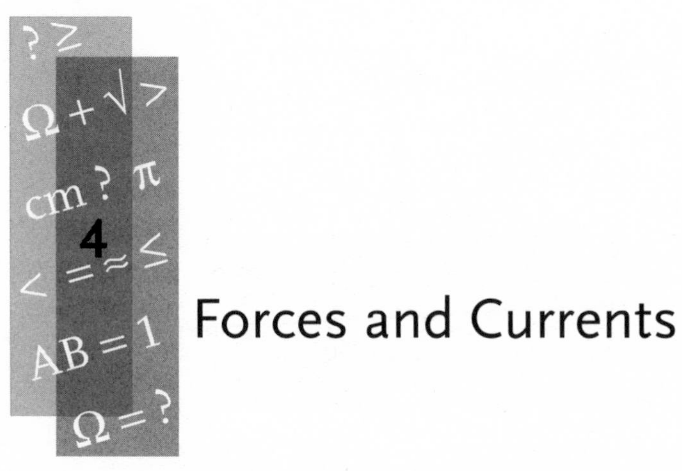

4 Forces and Currents

This chapter contains problems that assume familiarity with mechanics and electricity at the level of introductory college courses, but several of them may challenge even those with a deeper knowledge of physics.

PROBLEMS

4.1 Separated Boxes

Two equal boxes (A and B) stand on a floor with two sides parallel and only a few centimeters apart (fig. 4.1). Insert a long stiff plank vertically in the space between the boxes and pull horizontally at the upper end of the plank, in an attempt to separate the boxes. Will both boxes move, or only one of them?

4.2 Dropped Books

Take a stack of nine identical books and hold them horizontally by pressing them together with your hands (fig. 4.2). Then decrease the pressure slowly until the books are just about to start falling down. Which book(s) will start to slide first?

4.3 The Egg of Columbus

Christopher Columbus is said to have been at a dinner in Spain, where he asked if his host could make an egg stand on one end,

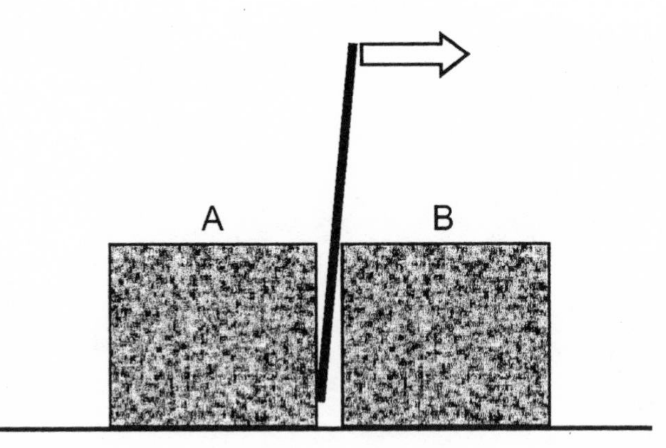

Fig. 4.1. Separating two boxes with a plank. Will both boxes move?

when placed on the marble table. When the host failed, Columbus took the egg, and broke the shell a bit so that it would stand upright. A seemingly difficult problem might be solved by anyone, after one has been shown how to do it, was Columbus's remark. The story of the egg of Columbus seems to be connected originally with Filippo Brunelleschi, who constructed the remarkable dome in the cathedral of Florence. It is said that when Brunelleschi met with opposition against his brave and ingenious idea, he suggested that any rival who could make an egg stand upright on the table should be given the task to finish the building of the cathedral. When they failed, Brunelleschi solved the egg problem in the way that has later been associated with Columbus.

The problem we shall now address also deals with balancing eggs, and has a simple suggested solution. Try to balance one egg on another egg, which rests on a table. You do it by taking an empty

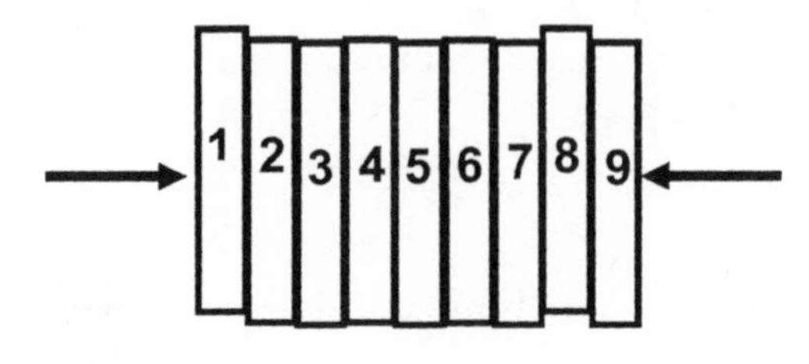

Fig. 4.2. Hold nine books horizontally between your hands. Loosen the grip somewhat. Which book(s) will tend to slide down first?

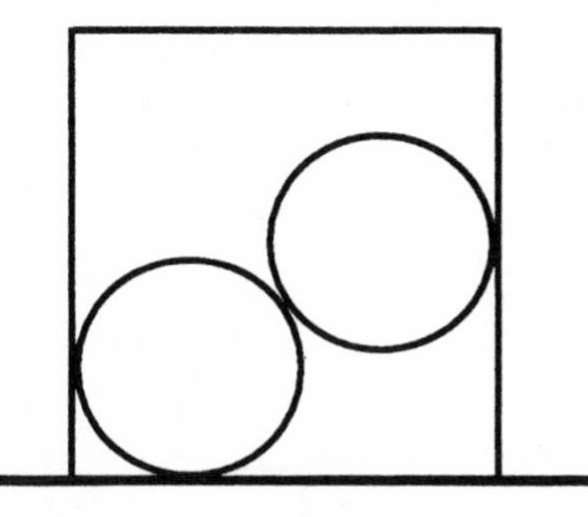

Fig. 4.3. Schematic illustration of two eggs inside an inverted plastic cup that has been placed on a table

cylindrical plastic cup, which originally contained 500 grams of yogurt. The cup is inverted and put over the two eggs, so that the inner wall of the cup prevents the upper egg from falling down on the table (fig. 4.3). Does it work if you are not holding the cup, or will the cup tumble over, with the result that both eggs lay on the table?

4.4 Helium or Hydrogen in the Balloon?

It has been suggested that balloons may be used to move very heavy objects. The lightest gas, with which the balloons could be filled, is hydrogen (H_2), but for safety reasons helium (He) is preferred. H_2 has the relative molecular mass 2. The corresponding value for He is 4, that is, it is twice as heavy. How much smaller (in per cent) is the load that could be lifted by the balloon, if hydrogen is replaced by helium. Ignore the mass of the balloon itself.

4.5 Lightbulb Found in a Drug Store?

While sitting at the kitchen table, you are given an incandescent lightbulb. It fits the standard U.S. socket size, but its power rating cannot be seen. The only equipment you have available is an ordinary ohmmeter. With that you measure the resistance of the lightbulb as 16 Ω. Is it likely that such a lightbulb can be bought in an ordinary U.S. supermarket?

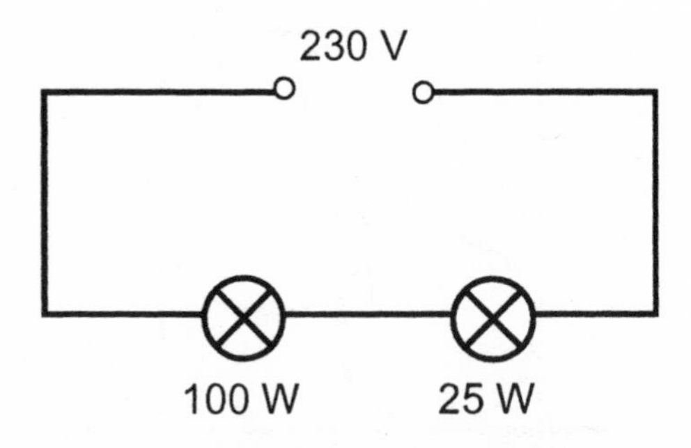

Fig. 4.4. Two incandescent lightbulbs, marked 100 W and 25 W, are coupled in a series. Which bulb is brightest?

4.6 Bright or Dark?

Two ordinary incandescent lightbulbs, bought in Europe, are marked 230 V, 100 W and 230 V, 25 W, respectively. They are connected in a series to 230 V (fig. 4.4). Will both, none, or only one of them light? If only one lights, which one will it be?

4.7 Yin and Yang

The well-known yin and yang symbol is often said to represent balance and harmony. Inspired by its geometry we consider a problem of "balance." A thin form has a shape that is similar to one half of the yin-yang symbol; see the filled area in figure 4.5. The form takes the orientation shown in the figure if it is suspended on a thin thread, at the distance x from the vertical line through the symmetry center of the form. What is the value of x? The shape of the form is defined by circular arcs as indicated in figure 4.6.

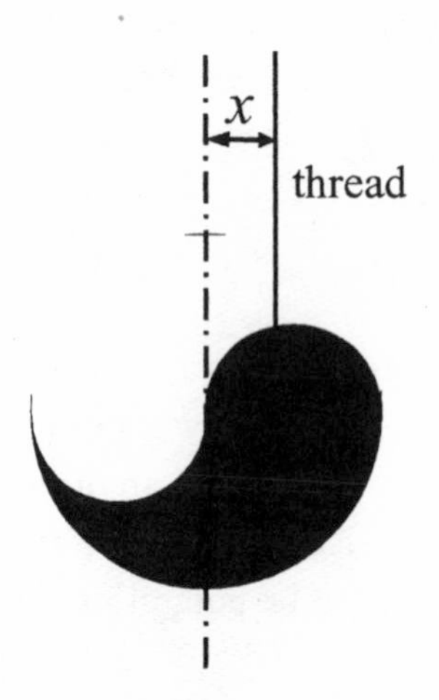

Fig. 4.5. Black form hangs suspended on a thread. What is the value of *x*?

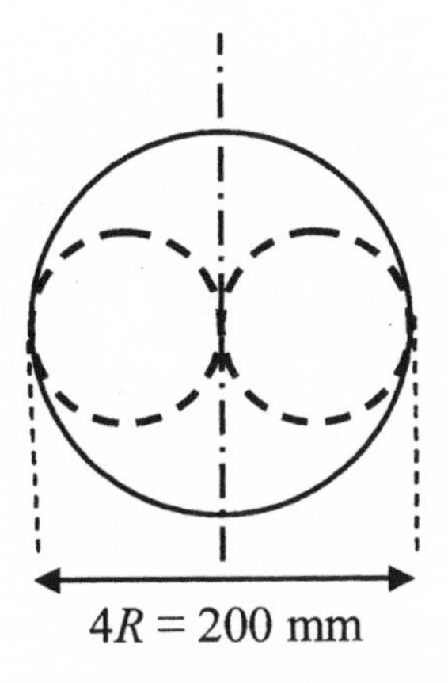

Fig. 4.6. Black form in figure 4.5 is bounded by circular arcs.

4.8 Rise and Fall of a Ball

A tennis ball is thrown vertically up and returns to the thrower. Which takes the longest time, the ascent or the descent?

4.9 Elevator Accident

In a high-rise building there is an elevator that has the same speed on its way up as on its way down. Suddenly a brick falls from the top of the shaft, down onto the elevator. Is the risk of damage to the elevator largest if the elevator is going down or if it is going up?

SOLUTIONS

4.1 Separated Boxes

Only the box to the right in the figure will move. The lower end of the plank tries to push box A to the left. As a reaction, a force of the same magnitude but in the opposite direction acts on the plank. In an analogous way, the plank tries to push box B to the right with a certain force, resulting in a reaction force on the plank. There is no net force on the plank. (Otherwise the plank would accelerate to the left or to the right. We ignore the weight of the plank itself and vertical frictional forces.) With forces as in figure 4.7, we must have

$$F_2 = F + F_1$$

Therefore

$$F_2 > F_1$$

Box B is pushed sideways by a larger force than the force acting on box A. Since the boxes were assumed to be equal, the maximum frictional force between box and floor is the same for both boxes, and the box to the right will start to move. You may try this problem on a small scale, for instance, by using a ruler and two equal cardboard packages of lump sugar on your kitchen table.

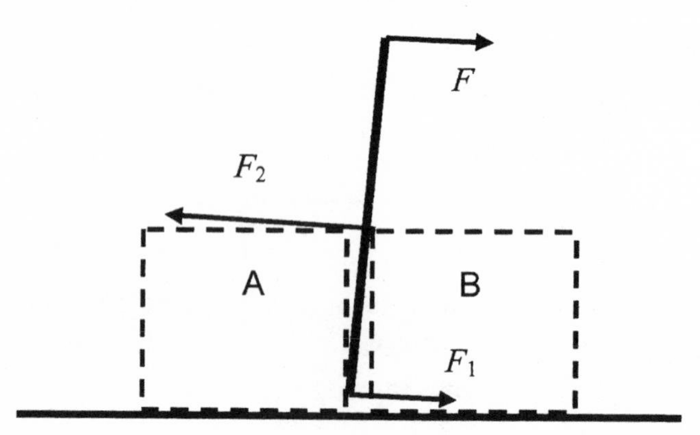

Fig. 4.7. Forces acting on the plank

Any caveat? As the separation between the boxes increases we get the (exaggerated) geometry in figure 4.8. The force from the plank has a component that presses down on box B and therefore increases

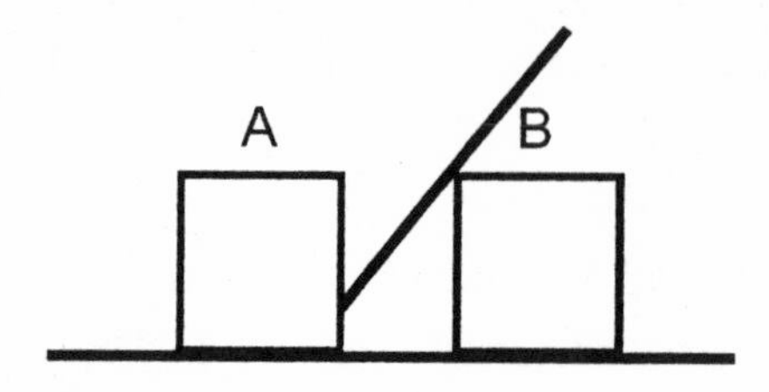

Fig. 4.8. When the boxes are widely separated, box A tends to be lifted, and box B to be pressed down. That changes the frictional forces between the boxes and the floor.

the frictional force on that box. Similarly, a force from the plank tries to lift box A, thus decreasing the frictional force on that box.

4.2 Dropped Books

All books except the two at the ends are likely to slide down together.

Simplify the problem and let there be only five books, all with mass m (fig. 4.9). First, consider the book in the middle. A total frictional force mg is needed to prevent it from sliding down. Then consider the three books in the middle, as one unit. (For instance, imagine that they are glued together.) The frictional force $3mg$ is needed to prevent these three books from sliding down. Finally, consider all five books as one unit. The frictional force $5mg$ is needed to prevent them from sliding down together.

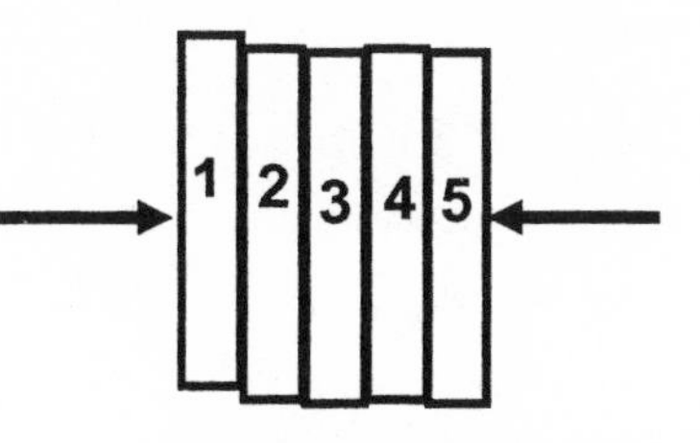

Fig. 4.9. Special case with only five books

Let the static friction factor between two books be f. The normal force N acting at each interface between the books is equal to the horizontal force F from one hand on the end of the book pile. The maximum vertical force $fN = fF$, therefore, is the same at all interfaces between the books. In our simplified case with only five books, it is now clear that the three books in the middle (with the weight $3mg$) will tend to slide as one unit before the single book in the middle (with weight mg) starts to slide. The possibility remains that all five books slide down together, which can happen if the friction factor f_{hand} between the hand and a book is too small. This normally is not the case, and we conclude that the three books in the middle slide down together. In the original problem we considered nine books. The same reasoning as above shows that the seven books in the middle slide together, when the grip from the hands is slowly loosened.

We now quantify the condition that there is no sliding between the hands and the outermost books. To be more precise, consider r books and find the condition that the $r-2$ books in the middle, rather than all r books, slide down. That requires the following inequalities to be fulfilled:

$$2f_{\text{hand}}F > rmg \text{ (no sliding between hand and book)}$$

$$(r-2)mg > 2fF \text{ (sliding of book against book)}$$

We then get

$$f_{\text{hand}} > [r/(r-2)]f$$

For $r = 9$ this means that the friction factor f_{hand} between hand and book should be about a factor of 9/7, or 30 %, larger than the friction factor f between two books. This is usually the case.

How physicists think. One can often avoid making trivial, or even fundamental, errors if a simpler case is first investigated. But it is important not to throw out the baby with the bath water. We started with a discussion of five books because that contains all the essential aspects of the complete problem with nine books. Taking only three books would leave out the possibility that several books in the middle slide together, but suffices to compare the sliding between book and hand, with the sliding between books.

Although an experienced physicist has no doubt about what is the normal force acting at the interfaces between the books, someone who is new to the field might feel uncertain and wonder if the normal force should be $N = 2F$ (rather than $N = F$) because we press on a certain book from *both* sides. A useful technique, in a case like this one, is to compare with another equivalent case, where you are sure of the answer. For instance, consider a light book lying on a table. Press down on the book with a force F. The normal force N from the table on the book (or from the book on the table), that is, the force to be used in the expression fN for the largest possible friction between the book and the table is fF. Replacing the table with your hand, and turning the system 90°, takes you to the geometry of the original problem. Some people may find this particular argument artificial or confusing. It is only meant to illustrate a technique that is of value in more than physics. In a case where you feel uncertain it can be very helpful to consider a different, but basically equivalent, case where you are more confident.

4.3 The Egg of Columbus

Normally the cup will tumble over.

Consider a simplified geometry. Let the two eggs be spherical, with radius r and mass m. The cylindrical cup has radius R and mass M. The eggs are placed inside the cup as shown in figure 4.10. It also shows the three forces acting on the *upper* egg, that is, a normal force N from the inner wall of the cup, the weight mg of the egg, and the normal force F from the contact with the lower egg. If the angle between F and the horizontal is α, equilibrium in the horizontal and the vertical directions requires

$$N = F \cos \alpha$$

$$mg = F \sin \alpha$$

Eliminating F we get

$$N = mg(\cos \alpha / \sin \alpha)$$

The upper egg pushes on the lower egg with a horizontal force of magnitude $F \cos \alpha = N$. To keep the lower egg in equilibrium, the left wall must act on it with a balancing horizontal force of the same magnitude.

Next consider only three forces acting on the *cup*, that is, the cup's weight Mg and the two horizontal forces of magnitude N from the

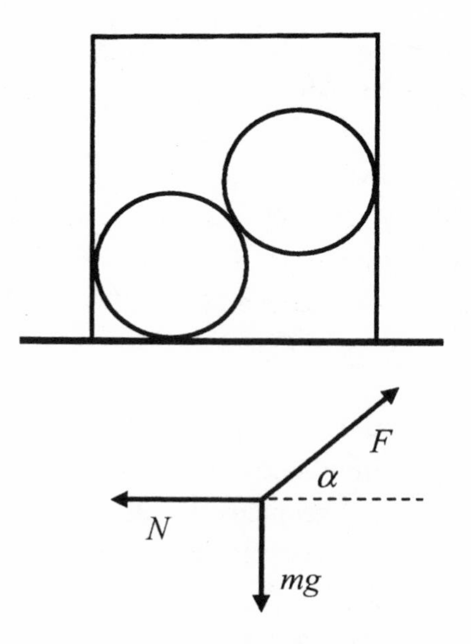

Fig. 4.10. Forces acting on the upper egg

two eggs. There are also forces from the table acting on the rim of the cup, but we are interested in the case when the cup tilts around A. With notation as in figure 4.11 we get a clockwise torque around A that is

$$Nb - Na - MgR$$

Furthermore, the geometry implies

$$a = r$$

$$b = r + 2r \sin \alpha$$

$$2r \cos \alpha + 2r = 2R$$

Putting all this together we get the clockwise torque around A

$$mg(\cos\alpha/\sin\alpha)(2r\sin\alpha) - MgR = 2mgR - 2mgr - MgR$$

If this expression is positive, the cup will tilt clockwise. It happens when the mass M of the plastic cup is sufficiently small compared with the total mass $2m$ of the eggs, that is, when

$$M < 2m(1 - r/R)$$

This is normally the case.

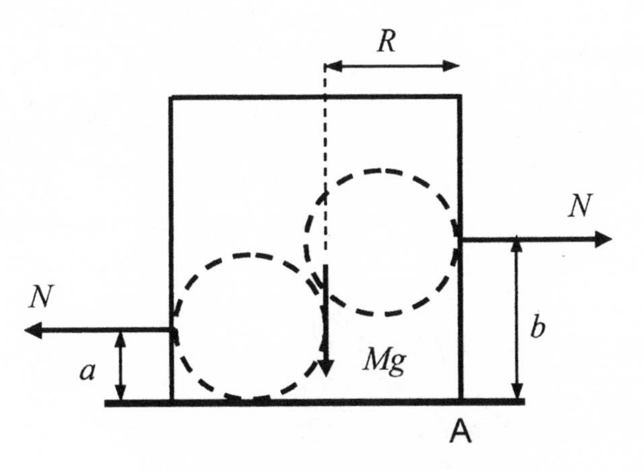

Fig. 4.11. Forces acting on the cylinder

How physicists think. Considering extreme cases may give a useful partial check of results. Let us therefore return to the condition for the cup to topple over:

$$M < 2m(1 - r/R)$$

We concluded that if M is small enough, the inequality is obeyed. But if $r > R$, the right-hand side is negative, and the left-hand side (the mass of the plastic cup) is, of course, positive. One should always check that such an apparent inconsistency has a logical interpretation and is not just the consequence of, for instance, an error in the calculation. In our case, $r > R$ means that a single egg is so large that it does not fit into the cylindrical cup.

Any caveat? We first check that the result is reasonable. A typical egg has the mass 60 g. Because the density of an egg, and the density of yogurt, are approximately the same as that of water, a cup that can hold 500 g of yogurt has about the same volume as eight eggs. That makes r/R of the order of 0.57 for a realistic shape of the cup. Furthermore, it is likely that the mass of the plastic cup is small enough compared with that of an egg. Thus, we have no reason to doubt the assumptions about mass and geometry, but it is possible that the forces are not correctly accounted for. Sometimes the cup will tilt a bit but not topple over, because the force geometry changes during the tilt and there is friction between the eggs and the inner walls of the cup. A calculation always rests on a simplified model. Basically physics is an experimental science. If an experiment can be performed, it may overrule the result of any theoretical model. But, of course, no experiment is perfect either.

4.4 Helium or Hydrogen in the Balloon?

The load is smaller by 7 %.

You may find it counterintuitive that filling a balloon with a gas that is *twice* as heavy reduces the possible load by as little as 7 %. Archimedes' principle says that the buoyancy force is equal to the weight of the air displaced by the balloon. The average relative molecular mass of air is about 29 (air has approximately 80 % of N_2 with relative molecular mass 28 and 20 % of O_2 with relative molecular mass 32). When the balloon is filled with hydrogen, and similarly when it is filled with helium, the buoyancy force must lift not only the payload but also the weight of the filling gas. For instance, if the displaced air has the mass 29 kg, the balloon contains 4 kg of helium gas or 2 kg of hydrogen gas, because the relative atomic mass of He is 4 and the relative molecular mass of H_2 is 2. The payload ratio therefore is

$$\frac{29-4}{29-2} = 0.93$$

The gas pressue in the balloon is usually close to that of the atmosphere, but our result would not be changed much even in a so-called superpressure balloon.

The air displaced by the payload itself also contributes to the buoyancy, but that is an extremely small correction in our case. For instance, if the payload is an iron construction with mass 8000 kg (17 600 lb), its volume is about 1 m^3 (35 ft^3). The mass of 1 m^3 of air is about 1 kg, and that is also the small increase in the payload.

Is it realistic? How big is a hydrogen or helium balloon that can lift, say, 8000 kg? It must displace air with a mass that is somewhat larger than 8000 kg. Let the density of air be 1 kg/m^3. Such a balloon has the same volume as a cube with sides 20 m (65 ft). It is big but not at all unrealistic.

A hot-air balloon must be bigger than a hydrogen or helium balloon to carry the same load. Material properties typically limit the temperature inside a hot-air balloon to 120 °C (250 °F). The hot-air pressure is approximately the same as the atmospheric pressure, because the balloon is open at its lower end and no significant pressure difference can be maintained there. The ideal gas law $pV = Nk_B T$, rewritten as the number of gas particles per volume $N/V = p/(k_B T)$, shows that the density varies as $1/T$. With an ambient temperature of 280 K and making the crude approximation that the temperature is 390 K everywhere inside the balloon, the density of the hot air is about $280/390 = 0.7$ of the density of the atmosphere. Since the mass of 1 m^3 of ambient air is about 1 kg, a hot-air balloon must envelope about 3 m^3 to lift 1 kg.

Outlook. Many remarkable events involve balloons. The first public ascents of a large, unmanned hot-air balloon, constructed by the French Montgolfier brothers, took place on 4 and 5 June 1783 (fig. 4.12). Soon after that, on 27 August 1783, the French physicist

Fig. 4.12. Classical hot-air balloon

Jacques Alexandre César Charles (whose name is remembered in the gas law) was the first to launch a hydrogen balloon. Benjamin Franklin described this event 3 days later in a long letter to the Royal Society in London. The disaster with the hydrogen-filled airship Hindenburg at Lakehurst, New Jersey, happened in 1937. The Hindenburg was actually built to be filled with helium, but U.S. restrictions on export of helium to Nazi Germany made this alternative impossible. Later, it was argued that it was the inflammable envelope, rather than the hydrogen, that was ignited by a spark from static electricity during the mooring. Large helium-filled balloons have been flown at least since 1921. In 1982 Larry Walters attached more than 40 helium filled weather balloons to his lawn chair and flew over Los Angeles.

4.5 Lightbulb Found in a Drugstore?

Yes, it is a very common type of lightbulb. When it is connected to 120 V, it has a power of about 60 W.

The equation

$$P = \frac{U^2}{R}$$

relating the power P to the resistance R and the voltage U may suggest that

$$P = \frac{120^2}{16}\text{ W} = 900\text{ W}$$

when the lightbulb is connected to the standard U.S. electrical voltage of 120 V. A lightbulb with such a high power rating is not found on the supermarket shelves. But this argument is misleading. The resistance measurement was made at the ambient temperature, say 295 K, while the bright lightbulb operates at almost ten times that temperature. The resistivity of tungsten, the material of choice for the filament, increases strongly with the temperature, like the resistivity of other metals. At operating temperatures the resistance has increased to about 240 Ω. The power then becomes

$$P = \frac{120^2}{240}\text{ W} = 60\text{ W}$$

Such a lightbulb is readily available.

Figure 4.13 shows how the resistivity of tungsten varies as a function of temperature. The resistance of the filament is about a factor of 15 higher when the lightbulb is lit, compared with the room temperature value. If we use room temperature data of $d\rho/dT$ in a linear extrapolation to estimate the resistance at the operating temperature, we get a value that is about 25 % too low.

Any caveat? The resistance at room temperature is measured with an instrument that uses direct current, while the lightbulb operates with 60 Hz alternating current. The filament is shaped as a coil, or even a “coiled coil,” and one may wonder if the inductance is small

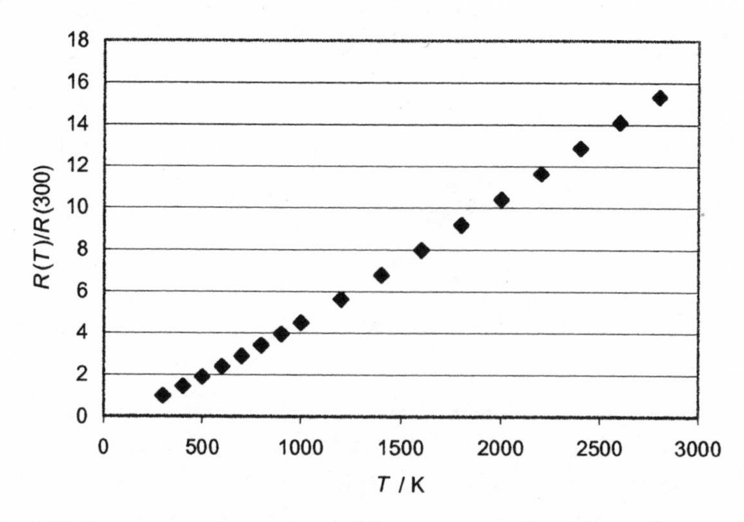

Fig. 4.13. Temperature dependence of the resistance of a tungsten filament, plotted as $R(T)/R(300\text{ K})$ versus T

in comparison with the resistance. A more detailed analysis shows that the inductance can be completely ignored in this context.

4.6 Bright or Dark?

Only the lightbulb marked 25 W will be bright.

Applying the equation

$$P = \frac{U^2}{R}$$

which relates the power P to the voltage U and the resistance R, would give the resistances 2100 Ω (25-W bulb) and 530 Ω (100-W bulb), respectively. Their resistances thus are in the ratio 4:1. When a potential U is applied to the lightbulbs coupled in the series, $0.8\,U$ falls over the 25-W lightbulb and $0.2\,U$ over the 100-W lightbulb, if we (incorrectly) assume that the filament temperature is the same for the two lightbulbs. Therefore, only the 25-W lightbulb will light up. This means that the temperature of its filament increases a lot, whereas that of the 100-W lightbulb remains rather low. The resistance ratio now becomes much higher than 4:1, which would be the ratio if the two filaments had equal temperatures. An even larger portion than 80 % of 230 V is therefore applied to the 25-W lightbulb. It will light with almost normal brightness, while the 100-W lightbulb remains dark.

4.7 Yin and Yang

The distance is 25 mm.

The black part takes the indicated stable orientation if its center of gravity lies vertically below the suspension point. We find the center of gravity as follows. Figure 4.14, with a gray and a black part, has its center of gravity on the vertical symmetry line. It can be considered as composed of two parts; the yin-yang form in our problem (black) and the circular disk (gray).

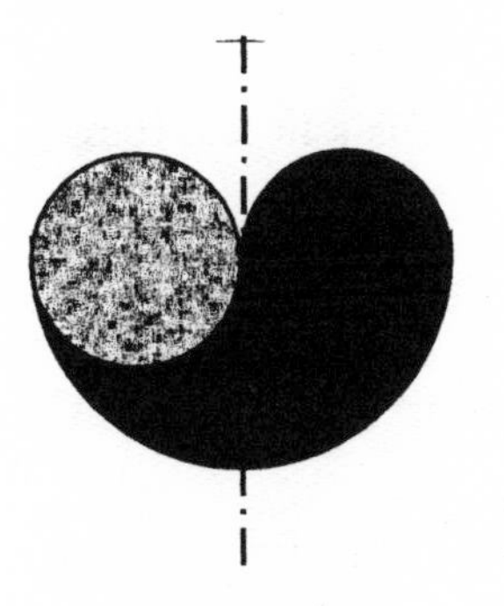

Fig. 4.14. The center of mass of this figure, composed of a circular disk and the black yin-yang form, is on the vertical symmetry line.

Let the small circular disk have radius R and mass m. The yin-yang-like black form then has a mass of $2m$ because its area is that of half a circular disk with radius $2R$. The combined parts have the center of gravity on the vertical symmetry line of the two parts, that is, the gravitational forces on the gray and the black parts balance:

$$mgR = (2m)gx$$

This condition gives

$$x = R/2 = 25 \text{ mm}$$

Fig. 4.15. The yin-yang symbol is often inscribed in a circle and has two small dots.

Finally, we note that the yin-yang symbol often is a little different from that in our problem (fig. 4.15).

4.8 Rise and Fall of a Ball

The descent takes the longest time, but the effect is so small that it may be difficult to notice in practice.

Without air resistance, the time would be the same for the ascent and the descent. Although the speed decreases with the height h above the ground, it is independent of the direction of the motion at that particular height when air resistance is ignored. But with air resistance present, energy is continuously dissipated to the surroundings. At every height, the speed is lower when the ball is on its way down. The descent therefore is slower.

Does it matter? Perhaps the effect is negligible for any reasonable speed we can give the ball? Often it is a good approximation to write the retarding force F due to air resistance as

$$F = \tfrac{1}{2} C_D \rho A v^2$$

Here ρ is the density of air, A is the cross-section area of the ball, and v is its speed. The dimensionless drag coefficient C_D is of the order of 1 and expresses how "streamlined" the object is. The numerical value (½) of the prefactor is just a convention and of no concern here. (Another convention would change C_D correspondingly.) The total energy that is lost because of air resistance when the ball goes up from the level $x = 0$ to the maximum height $x = H$ is the force times distance, that is,

$$\Delta E = \int_0^H F(x)\mathrm{d}x$$

Here $F(x)$ is the force calculated from $F = \frac{1}{2}C_D \rho A v^2$, with a speed v that depends on the height x. To get a feeling for the role of air resistance we assume that ΔE is only a *small* correction to the total energy of the system. To be more precise, we assume that v^2 in the expression for F can be replaced by its value when there is no air resistance at all;

$$\frac{1}{2}mv^2 = E_0 - mgx$$

The symbols m and g have their usual meanings and E_0 is the energy of the ball when it is released. Then we get

$$\Delta E = \frac{1}{2}C_D \rho A(v_0^2 H - gH^2) = \frac{1}{2}C_D \rho A g H^2$$

Here we have used that $\frac{1}{2}mv_0{}^2 = mgH$, when v_0 is the release speed at $x = 0$.

A *small* effect of air resistance means that we are comparing it with something larger. It is then a good start to consider the ratio $\Delta E / E_0$, that is, how much of the initial energy that is dissipated. We get

$$\frac{\Delta E}{E_0} = \frac{\frac{1}{2}C_D \rho A g H^2}{mgH} = \frac{\frac{1}{2}C_D \rho A H}{m}$$

According to the tennis rules, $m \approx 0.058$ kg. The ball's area is $A \approx 33$ cm^2. The density of air depends on temperature and pressure and is approximately 1.2 kg/m^3. As an illustration we now calculate the height H, which would give $\Delta E / E_0 = 0.1$. If we crudely take $\frac{1}{2} C_D \approx 1$ the result is $H = 1.5$ m. That seems to be such a remarkably small value that we should take a second look. Could there really be a large effect on the time to return, when the ball's maximum height is only 1.5 m (5 ft)? No, that is not true, as we will now see.

We took $\frac{1}{2} C_D \approx 1$, but this is too large. A better value would be $\frac{1}{2} C_D \approx \frac{1}{4}$, which increases the estimated H to 6 m. There is also another important aspect that we can qualitatively describe as follows. Assume that a ball is dropped to the ground from a certain height. In the absence of air resistance this would take the time t_0. With air resistance taken into account the ball has not yet reached the ground after the time t_0. It remains a certain distance, but the time to cover that distance is short compared with t_0 because the ball has a high speed when it is close to the ground. Phrased in other words—during *most* of the time it takes a ball to fall a certain distance, its speed, and therefore also the air resistance, is rather low. As a numerical example, let the ball reach the height $H = 5$ m. The total time for the path up and down is about $(1 + 1)$ s $= 2$ s. (The well-known relation $s = gt^2/2$, with $s = H = 5$ m, gives $t = 1.0$ s.) We will later see explicitly that the time delay is exaggerated if we consider H obtained from $\Delta E / E_0$. In the light of this discussion, it will be difficult to measure the duration of the throw, for instance, with a stopwatch, so accurately that the effect of air resistance is noted.

There is a closed form of the solution to *Newton's equation of motion* when an object with mass m falls under the influence of the gravitational force mg and the retarding force $F = \frac{1}{2} C_D \rho A v^2$ from air resistance. If the motion starts from rest, the distance s covered after the time t is given by

$$s = a \ln\left[\cosh\left(t\sqrt{\frac{g}{a}}\right)\right]$$

where a is a parameter with the dimension of length,

$$a = \frac{2m}{C_D A \rho}$$

The expression for s is here given without derivation, but we may check it by forming the acceleration $\mathrm{d}^2 s/\mathrm{d}t^2$. If we combine it with the speed $v = \mathrm{d}s/\mathrm{d}t$ we get

$$\frac{\mathrm{d}^2 s}{\mathrm{d}t^2} = g - \frac{F}{m}$$

as expected.

Using the series expansions $\cosh(x) = 1 + x^2/2 + x^4/24 + \ldots$, $\ln(1 + x) = x - x^2/2 + \ldots$, and $(1 + x)^{1/2} = 1 + x/2 - x^2/8 + \ldots$ one finds, after some manipulations, that inclusion of the lowest-order correction due to air resistance (i.e., large a) gives a time t to fall a distance s from rest as

$$t = \sqrt{\frac{2s}{g}}\left(1 + \frac{s}{6a}\right) = \sqrt{\frac{2s}{g}}\left(1 + \frac{C_D}{12}\frac{M_{\text{air}}}{m}\right).$$

In the last equality we introduced the quantity

$$M_{\text{air}} = sA\rho.$$

M_{air} is the mass of the air in the "tube" of cross-section A and length s that is swept out by the falling object.

How physicists think. This problem contains many examples of how physicists think. Instead of performing a detailed calculation, which could also be done numerically, we first identified several important aspects. The effect of air resistance is certainly present to slow down *all* throws. We found, however, that it would not be noticeable in practice, and there is no reason to pursue the detailed analysis. There are still a few aspects worth discussing. The treatment in the following paragraph might seem a bit dry, but it contains some ideas really characteristic of physics thinking, and it can be rewarding to pay attention to its details.

Rather than trying to solve the equation of motion analytically, we first assumed that the speed v was *unaffected* by the air resistance. That speed was then used in the expression for the air resistance, $F = \frac{1}{2}C_D \rho A v^2$. From a strictly logical point of view, this is certainly inconsistent. But physics deals with real situations, described by models that are only approximate. Then our approach is not only adequate for the present problem of the tennis ball, but it represents a very important method in physics—*perturbation theory*. Suppose, as an example, that the air resistance gives a small effect that would change the speed v by 1 %. We can call that change a perturbation. When the air resistance force F is calculated as $\frac{1}{2}C_D \rho A v^2$, the 1 % change in v results in a 2 % change in F (because v appears squared in F). But air resistance is a very complicated phenomenon, and $\frac{1}{2}C_D \rho A v^2$ does not give the retarding force for a tennis ball as accurately as 2 %. Therefore, it makes no difference in practice if we take a speed v that is uncorrected for the air resistance, when we put this v into the expression for F. Of course, this approach only works when the correction (perturbation) is in some sense small. Exactly what is meant by "small" depends on how accurate we want the analysis to be.

There is an interesting interpretation of the correction term

$$\frac{s}{6a} = \frac{C_D}{12} \cdot \frac{M_{air}}{m}$$

in the expression for the time t above, and also in the related ratio

$$\frac{\Delta E}{E_0} = \frac{\frac{1}{2}C_D \rho A H}{m} = \frac{1}{2}C_D \cdot \frac{M_{air}}{m}$$

obtained with $s = H$. If the mass of the air in the tube that is swept out by the tennis ball is much smaller than the mass of the tennis ball, that is, $M_{air} << m$, we also have that the correction terms $s/(6a) << 1$ and $\Delta E/E_0 << 1$. The condition $M_{air} << m$ therefore is another way of expressing the intuitively obvious fact that a heavy ball would be less affected by the air resistance than a light ball. But a word like "heavy" has a meaning only if one *compares* things with the dimension of mass. A heavy man has a mass that is at least larger than the average mass of a man. A heavy atom may have a mass that is larger than the masses of other atoms of interest. In our case, a heavy ball has a mass that is larger than the mass of the atmosphere inside the tube cut out by the ball along its entire trajectory.

A further twist of the same issue is to say that if the distance H of the fall obeys the inequality

$$H << \frac{2m}{C_D A \rho} = a$$

then the air resistance has a negligible effect on the time to complete the fall. The quantity a (which we introduced earlier in this problem) is now clearly identified as a *characteristic distance* of relevance when we model the motion of a tennis ball in air. It is of a universal nature, that is, it does not depend on the specific situation regarding release speed, release angle, and so on. With data for air and for the tennis ball, and with $C_D = 0.5$, we get

$$a \approx 60\ m.$$

In our case, air resistance is small for all heights $H << 60$ m. A physicist who is faced with a complex problem will ask for characteristic quantities, like times, masses, speeds, and so on. They will help to sort out relevant factors from those that can be ignored.

Any caveat? The expression for the air resistance,

$$F = \tfrac{1}{2} C_D \rho A v^2$$

suggests that F varies as the square of the speed v. That is only an approximation, because a further dependence on speed is hidden also in the dimensionless quantity C_D. We can write $C_D(Re)$ for a spherical object as a function of the dimensionless *Reynolds number*

$$Re = \frac{vD\rho}{\eta}$$

Here η is the dynamic viscosity of the air and D is the radius of the sphere. In air (20 °C) $Re = 7 \times 10^4\, vD$ when v is expressed in meters per second and D is in meters. As an example, a tennis ball with $D = 0.065$ m, thrown with the speed $v = 10$ m/s, has $Re = 5 \times 10^4$. $C_D(Re)$ for a smooth sphere is approximately constant in a very wide range, $10^3 < Re < 2 \times 10^5$. Then it is a good approximation to ignore the dependence on speed that is hidden in C_D, and let F vary as v^2. The precise value of C_D depends on how smooth is the surface of the sphere and therefore may be different for a new and a worn tennis ball. Measurements have shown that $C_D \approx 0.6$ for tennis balls.

With the increasing speed of a falling object, the air resistance grows until eventually it is equal to the force of gravity on the object. Then the object is no longer accelerated but falls with a constant *terminal speed* (terminal velocity). Here are some examples. The terminal speed is approximately 50 m/s (180 km/h or about 110 mile/h) for an outstretched human, about 22 m/s for a tennis ball, and about 8 m/s for a table tennis ball. Without air resistance a tennis ball would reach the speed of 22 m/s after 25 m (82 ft) of free fall. Raindrops rapidly reach their terminal speed; about 6.5 m/s for a typical raindrop with a diameter of 2 mm. With this speed and size, the Reynolds number of a raindrop is about 900, that is, a bit too small for the ordinary air resistance relation to be accurate. In clouds the droplet diameter is about 0.02 mm and the terminal speed is 0.01 m/s. In that case the retarding force comes from the viscosity of the air, and the relation $F = \frac{1}{2} C_D \rho A v^2$ cannot be used at all. In some parts of the world it is common to celebrate by shooting vertically into the air. The returning bullet has a terminal speed that is usually too low to kill a person. According to a popular myth, a penny thrown from the Empire State Building would embed itself into the street and could even kill a person that is hit. This is not true. Apart from the fact that winds around the building normally sweep the penny away to land some stories below, a coin reaching the street level is not likely to cause much harm.

Before closing the discussion on the fall of a ball, we return to the concept of characteristic quantities. The terminal speed $v = v_{term}$ is that speed at which the air resistance balances the gravitational force mg. We have

$$\frac{1}{2} C_D \rho A v_{term}^2 = mg$$

We can also define a speed v_H as that which a mass has reached after freely falling the distance H. In a fall where

$$v_H << v_{term}$$

we can ignore the air resistance. This condition can also be written

$$\sqrt{\frac{C_D \rho A H}{m}} << 1$$

which is qualitatively the same as we found earlier through another argument. Instead of comparing H with a characteristic distance a, we can compare v_H with a characteristic speed v_{term}.

4.9 Elevator Accident

It makes no difference. The relative impact speed is the same.

We ignore the effect of air resistance on the brick. The damage depends on the relative speed of the elevator and the brick at the impact. When the elevator is on its way up, we should add the speeds of the brick and the elevator, while we should take their difference when the elevator is going down. As we shall see, the relative impact speed is the same in the two cases.

Let the elevator be the distance h below the brick, when the brick starts to fall. First, we assume that the elevator is going up with the speed v (fig. 4.16). It takes the time t before the brick hits. In its free fall it has reached the speed

$$u = gt$$

and it has fallen the distance

$$s = \frac{1}{2}gt^2$$

Furthermore the sum of the distances covered by the brick and the elevator during the time t is equal to h, or

$$h = s + vt$$

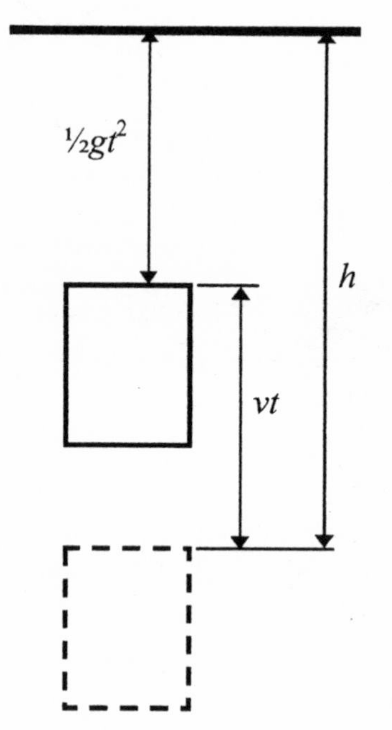

Fig. 4.16. Originally the distance h below the top, the elevator ascends the distance vt. The brick falls the distance $\frac{1}{2}gt^2$.

Inserting $s = \frac{1}{2}gt^2$ we get the quadratic equation

$$h = \frac{1}{2}gt^2 + vt$$

or

$$t^2 + \frac{2v}{g}t - \frac{2h}{g} = 0$$

with the solution (the other root gives a negative time t and therefore is unphysical)

$$t = -\frac{v}{g} + \sqrt{\frac{v^2}{g^2} + \frac{2h}{g}}$$

We can now express the relative speed v_{rel} of the brick and the elevator at the impact as

$$v_{rel} = v + u = v + gt = \sqrt{v^2 + 2gh}$$

If the elevator is going down, we should replace v by $-v$. That does not change v_{rel}, and we conclude that the damage to the elevator is the same when it goes up as when it goes down.

How physicists think. Even though the calculations are simple, it is a good practice to verify that the result is correct in special cases—in particular, because it may seem a bit surprising that the risk of damage is independent of the direction in which the elevator moves. The first, and trivial, step would be to check that the expression

$$v_{rel} = \sqrt{v^2 + 2gh}$$

is dimensionally correct. Next, we assume that the elevator is on its way up and is close to the top ($h \approx 0$). Then the brick has not had time to speed up before the impact, and the relative speed is just the speed v of the elevator. This agrees with our result for v_{rel}. If the elevator is on its way down ($v < 0$), the relative speed at impact is

$$v_{rel} = u - |v|$$

In the limit that $h \approx 0$, this means that the speed of the brick, $u = v_{rel} + |v|$, is twice that of the elevator at impact. Checking it from the relations shown above is like arguing in a circle, and it is much better to do a separate calculation from scratch (not shown here). Finally, when the elevator is not moving we have $v = 0$, which gives

$$v_{rel} = \sqrt{2gh}$$

As expected, this result agrees with the speed the brick has after falling the distance h.

Does it matter? Different elevators have different speeds v. We therefore investigate how sensitive the result is to variations in v. It was found that

$$v_{rel} = \sqrt{v^2 + 2gh}$$

If $v^2 << 2gh$, it is the *height* h that matters for the relative speed v_{rel} at impact. When $v^2 >> 2gh$, it is the *speed of the elevator* that is the crucial quantity. The "borderline" between these two cases is given by the relation

$$v^2 = 2gh$$

As an example, we first consider the speed $v = 1$ m/s, which could be typical of elevators in low buildings. The borderline condition implies that $h = v^2/(2g) = 5$ cm (2 in). Next we take $v = 7$ m/s, which is a speed found only in some high-rise buildings (cf. table 3.1 in the solution to problem 3.10). The borderline distance is $h = 2.5$ m (8 ft). We conclude that according to our analysis the speed of the elevator is important only for distances h so small (<5 cm in the first case and <2.5 m in the second case) that the risk of damage is either entirely negligible or at least quite small. (The risk, of course, is considerable when the height h is large, irrespective of the elevator's speed.) The problem, with its counterintuitive result, turns out to be an amusing exercise rather than a realistic modeling of a serious accident. It would have been good "physics thinking" if we had made a simple estimation and reached this conclusion before we embarked on the detailed mathematics.

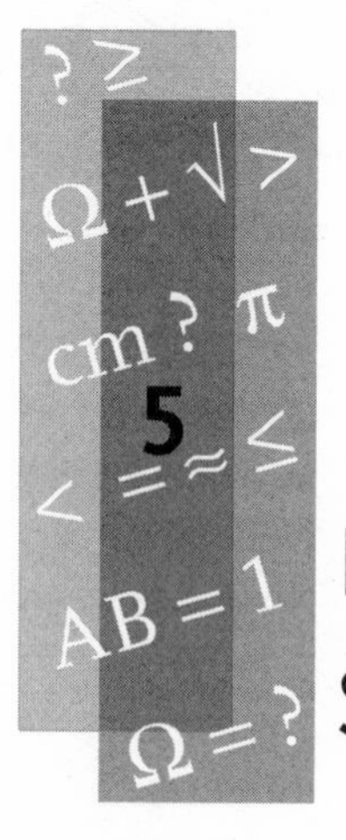

5 Not Exact but Still Relevant

Many realistic problems cannot be given an "exact" answer, but an approximate answer is good enough. That is the theme of this chapter.

PROBLEMS

5.1 What Is Your Volume?

People can immediately tell their own height and their own mass, but what about their own volume? The height of a typical adult woman in northern Europe is 165 cm (5 ft 5 in) and her mass is 65 kg (140 lb). What is the volume of that woman?

5.2 On the Move

Many airports with long distances to the gates have moving walkways, like conveyor belts for people. Your friend suggests an experiment. You are asked to walk from one end of the moving belt to the other, in the direction of the belt's motion, and then back again on *the same belt*. Your friend will walk on the corridor floor, to the end of the belt and back. Who returns first to the starting point?

5.3 Shot Put and Pole Vault

You are watching the Olympic athletic events shot put and pole vault. What is passing highest above the ground—the center of mass of the shot or the center of mass of the pole-vaulter?

5.4 Record Stadium

According to the rules for hammer throwing, issued by the International Association of Athletics Federations (IAAF), "in all throwing events, distances shall be recorded to the nearest 0.01 m below the distance measured if the distance is not a whole centimetre." Furthermore, the rules say that the field must not have a downward slope exceeding 1:1000 in the throwing direction. Would the maximum allowed slope improve the results by more than 1 cm?

5.5 Grains of Sand

Are there more grains of sand in a sand heap than there are atoms in a single grain of sand?

5.6 Cooling Coffee

When you buy your meal in the canteen you also buy a cup of coffee. It is very hot, and you want the coffee to have a more enjoyable temperature when you drink it after the meal. Since you have also taken some cartons of cold milk, you wonder if it is best to pour the milk into the coffee immediately, or when you are ready to drink it? What is the best way to cool the coffee?

5.7 Time for Contact

Two homogeneous brass spheres, each with the mass 1 kg, are suspended in 1-m-long wires (fig. 5.1). One of the spheres is moved 1 dm to the side and released, so that it hits the other sphere. The spheres are in contact for a very short time t_c. Is t_c shorter than 1 μs?

Some data for brass

Composition (mass)	63 % Cu, 37 % Zn
Density, ρ	8400 kg/m³
Elastic (Young's) modulus, E	105 GPa

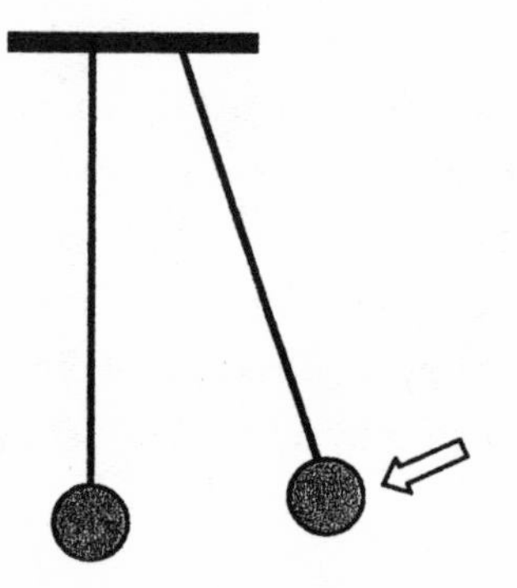

Fig. 5.1. What is the time of contact when two brass spheres collide?

5.8 Socrates' Blood

You drink a glass of ordinary tap water. How many water molecules in this glass were in the blood of Socrates when he died?

SOLUTIONS

5.1 What Is Your Volume?

The typical volume is about 65 liters, or 0.065 m^3.

One just barely floats in water, that is, the average density of the human body is a little less than that of water, or a little less than 1 kg/L = 1000 kg/m^3.

The human body consists of about 70 % of water. The rest has a higher density than water, leading to an average density that is about 1070 kg/m^3. This is consistent with the observation that when cooking a piece of meat, it does not float. But when we talk about the volume of a human body, we should also include the air-filled parts, in particular the lungs. Then one finds that the average density of a human body is about 950 kg/m^3 after inhaling and 1020 kg/m^3 after exhaling. Many people have experienced that it is a little easier to float in the salt water of the oceans than in a lake. The salinity gives the ocean water a density that varies somewhat with the geographical location and is about 3 % higher than that of pure water.

Of course it is not a coincidence that the density of water is so close to 1000 kg/m^3. When the base SI units for length (meter) and mass (kilogram) were introduced during the French Revolution in the eighteenth century, it was decided that 1 kg is the mass of 1 L (liter) = 1 dm^3 of water. However, such a relation is not accurate enough. The present definition, finally adopted in 1901 (and earlier in 1889), says that 1 kg is equal to the mass of the international prototype of the kilogram. It is a cylinder made of a platinum-iridium alloy, and kept in Sèvres outside Paris.

This problem shows how useful the SI units are. In the United States one may need to relate a length expressed in feet (1 ft = 0.3048 m, exactly) and inches (1 inch = 0.0254 m, exactly) to a mass expressed in pounds (1 lb = 0.453 592 37 kg, exactly), and then get a volume in cubic feet (1 cubic foot = 28.316 846 592 dm^3, exactly). In the British system one may prefer to give the mass in stone (1 stone = 14 lb; plural form, stone). Information on the size and mass of various parts of the human body is referred to as *anthropometric* data. They are important in many fields, from space flights to the design of clothing.

5.2 On the Move

Your friend will be first, even if you walk somewhat faster.

Before we try a mathematical solution, it is illuminating to consider an extreme case. Suppose that you walk on the belt with a speed that is *slower* than that of the belt itself. When you try to return, you are carried backward faster than you can walk forward on the belt. As a result you remain "stuck" at the end of the belt. If your speed is just *a little* bit faster than that of the belt, you are able to return to the starting point but it will take a long time. This argument gives the trend. It seems obvious that it is always slower to walk on the belt. A physicist might find this to be sufficient as an answer to the problem of who will return first, but here is a mathematical analysis.

Let the walking speed be u relative to the surface you step on (the corridor floor or the belt) and let the speed of the belt be v. The distance between the end points of the belt is L. The speed relative to the corridor floor is $u + v$ when you walk on the belt in its direction of motion, and it is $u - v$ when you walk in the opposite direction. The total time needed is

$$t = \frac{L}{u+v} + \frac{L}{u-v} = \frac{2Lu}{u^2 - v^2} = \frac{2L}{u} \cdot \frac{1}{1 - v^2/u^2}$$

The last factor is always larger than 1, so the time t is always larger than $2L/u$ if the belt is moving ($v > 0$).

Does it matter? Take the realistic values $L = 75$ m, $u = 1.5$ m/s, and $v = 1$ m/s. That gives $t = 100$ s for walking on the corridor floor, and 180 s for walking on the belt, that is, a very significant difference. To be more precise, let us calculate how much faster you must walk, compared with your friend, if you both return at the same time. Your speed is still u but your friend's speed on the corridor floor is w. You arrive back simultaneously if

$$\frac{2L}{w} = \frac{2L}{u} \cdot \frac{1}{1 - v^2/u^2}$$

which can be written as

$$\frac{w}{u} = 1 - \frac{v^2}{u^2}$$

A *very* fast walk corresponds to, say, $u = 2$ m/s. If 2 m/s is your speed and if the speed of the belt is still $v = 1$ m/s, your friend can walk on the corridor floor at the more modest pace of 1 m/s without returning later than you.

Note that the time t depends on the square of v. A "negative" speed would give the same time. Thus, it does not matter if you start walking in the same direction as the belt is moving and then return on the belt, or if you do it in the reverse order.

Our problem has a close analogy in one of the most famous experiments in physics—the Michelson–Morley experiment in 1887 to determine whether the speed of light depends on the speed of the light source through an assumed stationary ether.

5.3 Shot Put and Pole Vault

For the female athletes the shot passes approximately at the level of the bar in pole vaulting. For male athletes the shot does not quite reach the same height as the bar.

Recall a well-known result for the range s of a parabolic flight path, when the projectile is launched with speed v at an angle α to the horizontal ground. It is

$$s = \frac{v^2}{g}\sin(2\alpha)$$

where g is the acceleration of gravity. The maximum range s is obtained for $\alpha = 45°$. Then the maximum height of the trajectory is found to be

$$h = \frac{s}{4}$$

The best female and male shot putters reach about 21 m (69 ft). The shot does not leave from the ground level, however, but from a hand that is about 2 m higher up. If we still imagine that s corresponds to a trajectory starting at the ground, and at an angle $\alpha = 45°$, the shot has already covered a horizontal distance of about 2 m when it leaves the hand. That makes s longer by about 2 m than the result recorded at the competition. The expression $h = s/4$ then suggests that the shot passes almost 6 m above the ground. This is also about the height reached by the male pole-vaulter, but it is about 1.5 m higher than the result for the female pole-vaulter. Considering the uncertainties in our model it is best to conclude that the maximum height of the shot is somewhat lower than the level of the bar in pole vaulting for men, and about the same as that level for women. The winners' results at the 2004 Olympic Games in Athens were: pole vault, 4.91 m for women and 5.95 m for men; shot put, 21.06 m for women (shot mass, 4.00 kg) and 21.16 m for men (shot mass, 7.26 kg).

Any caveat? The release angle α affects the possible force on the shot, and therefore also its speed v. Since the aim is to maximize the actual $s(\alpha)$, we have $ds(\alpha)/d\alpha$ at the optimum α. This gives only a small (second-order) effect in s but affects the height in a complicated way. Many athletes let α be about 35°. Because $h/s = 1/4 \tan \alpha$ we have overestimated the height of the shot if we let $\alpha = 45°$.

The distance between the bar and the center of mass of the pole-vaulter is very small when the bar is passed. In fact, the center of mass often passes slightly below the bar, as is also the case in high jumping.

5.4 Record Stadium

The expected improvement is about 7–8 cm (3 in) when the field has its maximum allowed slope. The path of the hammer is close to that of a parabola with a launch angle of 45°. It will also land at approximately the same angle. The schematic illustration in figure 5.2 shows that the throw is approximately as much longer as the field at the landing point lies deeper. Since a very long hammer throw is about 80 m (men) or 70 m (women), we expect an improvement that is about 7–8 cm.

Any caveat? The hammer head is so small and heavy (men: diameter, 110–130 mm, and mass including grip and wire, 7.26 kg; women: diameter, 95–110 mm, and mass, 4.00 kg) that the aerodynamic effects are not important (unlike in discus and javelin throwing). With a release speed v of about 30 m/s, the air resistance F calculated from the expression $F = \frac{1}{2} C_D \rho A v^2$ is only about 5 % of the gravity

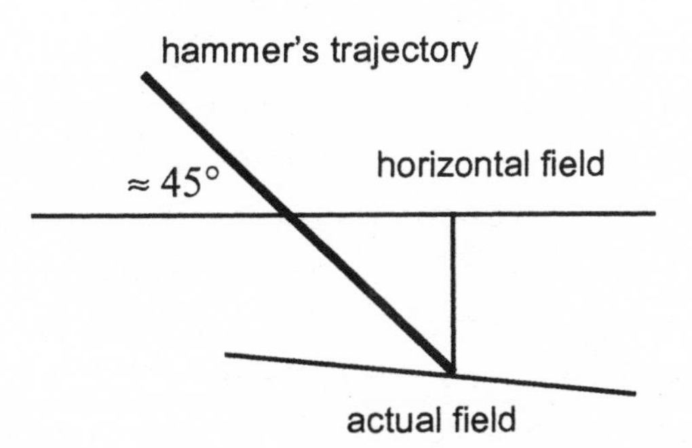

Fig. 5.2. The hammer lands at approximately 45° to the horizontal plane.

force mg on the hammer head. Instead of using the value of the release speed we could have used that the length of the throw is approximately $s = v^2/g$, which gives

$$\frac{F}{mg} = \frac{\frac{1}{2} C_D \rho A s}{m}$$

The numerator on the right-hand side can be interpreted as the mass of the air inside the tube cut out by the trajectory of the hammer head multiplied by $\frac{1}{2} C_D \sim \frac{1}{4}$. This aspect of air resistance is discussed in problem 4.8.

5.5 Grains of Sand

There are *many* more atoms in a single grain of sand than there are grains of sand in a sand heap.

This is not a precisely formulated problem, because the size of a sand heap is not well defined. The size of a grain of sand is also somewhat vaguely defined. Nevertheless, we will see that there is a definite answer. The result is so clear that we don't need to be very accurate in the estimation.

First, we recall some results from chemistry. One mole of an element contains about 6×10^{23} atoms (Avogadro's constant). The mass of one mole of a substance is "as many grams as the number which expresses the relative atomic mass." (The relative atomic mass was previously called the atomic weight.) Sand does not consist of a single chemical element, but for our purpose we can represent sand with an element having a relative atomic mass of, say, 30, and a mass density of 3 g/cm^3. One mole of sand then has the mass 30 g and the volume 10 $cm^3 = 10^4$ mm^3. Let a grain of sand have the form of a cube with sides of 1 mm, that is, the volume 1 mm^3. A single grain of sand contains about 10^{-4} mole of atoms, or about 10^{20} atoms. Is this number larger than the number of grains in a sand heap? The total volume of 10^{20} grains of sand, each with the volume 1 mm^3, would be 10^{11} m^3. That corresponds to the volume of a cube with sides significantly longer than 1 km (or 1 mile)—which would make a very large sand heap!

In *Metamorphoses,* written by the Roman poet Ovid 2000 years ago, Apollo asks Sibyl to choose whatever she wishes. She asks for as many years of life as there are grains of sand in her hand. But she refuses his offer of eternal youth in return for her eternal love, and she washes away until only her voice remains. We note that 1000 years of life corresponds to a few grams of sand, so Sibyl lived a very long life.

How physicists think. Obviously it was not necessary to worry about how large a sand heap is, what the chemical composition of sand is, or how large a grain of sand is. (Geologists usually call it "sand" when the grain diameter is from 0.2 to 2 mm. The precise definition is not the same in all countries.) But if the estimation had not given such a clear-cut result, we would have paid more attention to details.

This problem dealt with powers of ten. When they are explicitly written out, like 10^{12}, there is no ambiguity. But the number 10^{12} is often misinterpreted when given a name. In American English one says *trillion,* but in French the name is *billion* and in German it is *Billion,* with similar names in other European languages. The names of powers of ten in American English are given for the sequence

$$10^{3+3n}$$

as million ($n = 1$), billion ($n = 2$), trillion ($n = 3$), and so on. In the European languages the sequence is

$$10^{6n}$$

with names (in German, as an example) *Million* ($n = 1$), *Billion* ($n = 2$), *Trillion* ($n = 3$), and so on. The German word for the American English word billion is *Milliarde*. If you want to be absolutely unambiguous you can say "one thousand millions" instead of "one billion." Journalists who are unaware of these difficulties often make mistakes when translating from one language to another. A good training in estimations can be very valuable when one encounters data expressed in billions.

Additional challenge. What is the answer to our problem if "sand heap" is replaced by "sand desert"? (Solution at the end of this chapter.)

5.6 Cooling Coffee

You should normally wait to add the milk until you are ready to have your coffee. But it does not matter much!

This is a well-known problem that is often discussed incorrectly. Cooling is achieved by radiation, conduction, convection, and evaporation. Of these, radiation is usually the least important mechanism, but because it is sometimes the only one discussed we consider it first, and in some detail. Go to an extreme case and assume that the black coffee behaves like an ideal black body, whereas the "white" coffee (with milk) radiates like an ideal white (i.e., perfectly reflecting) body. *The Stefan-Boltzmann radiation law* says that the power P radiated from a black body of absolute temperature T and area A is

$$P = A\sigma T^4$$

where $\sigma = 5.670 \times 10^{-8}$ W/(m^2·K^4) is the Stefan-Boltzmann constant. A black body is not only perfect in emitting radiation, but also in absorbing the radiation that comes from its surroundings. If the coffee temperature is T_c and the ambient temperature is T_0, the net radiated power from a black surface is

$$P_{net} = A\sigma(T_c^4 - T_0^4) \approx 4A\sigma T_0^3(T_c - T_0)$$

Let the top area of the cup be $A = \pi r^2 = \pi \times 16$ cm$^2 = 5 \times 10^{-3}$ m^2, T_0 = 22 °C = 295 K, and $T_c - T_0 = 40$ K, giving $P = 1.2$ W of cooling power. In one minute (60 s), the heat loss is 72 Ws = 72 J ≈ 17 cal. That lowers the temperature of 170 g (6 oz) of water by 0.1 °C (0.2 °F). (Recall that changing the temperature of 1 g of water by 1 °C involves the energy 1 cal ≈ 4.2 J.) A coffee cup may hold somewhat more than 170 g, so the initial cooling rate would be less than 0.1 °C/min. That is certainly negligible. Furthermore the cooling rate decreases as $T_c - T_0$.

But is it true that white coffee radiates like a white body? The answer is no. A body that is white for light in a certain range of wavelengths may be black or grey at other wavelengths. Biological evolution has made our eyes sensitive to the "black body" radiation corresponding to the surface temperature of the sun, about 6000 K. Because we cannot see the infrared light that is emitted by bodies at ambient temperatures, we are not aware that "white" bodies may be "black" in the infrared. There is only a tiny difference in the radiation properties of white and black coffee. In both cases, the radiation from the liquid is well approximated by that from a black body. For water the emissivity is $\varepsilon \approx 0.95$ in the infrared. Not only water radiates almost like a black body. The same is true for snow. The "white" snow is actually "black"—in the infrared. Even white paint, the ordinary color of radiators for indoor heating, is black in the infrared. There is nothing to gain in painting them black, with the idea to make them smaller without diminishing the radiation part of the heat transfer to the room.

After we have found that cooling by radiation can be ignored, we look for other effects, which may depend on when the milk is added. Cooling is often described by *Newton's cooling law:*

$$\frac{dQ}{dt} = -k(T - T_0)$$

where Q is the heat transfer, t is the time, T is the temperature of the object that is cooled, T_0 is the ambient temperature, and k is a constant, which depends on the particular situation that is considered. One may remark that this differential equation is neither a law of physics, nor is it due to Newton. It has the solution

$$T(t) = T_0 + (T_i - T_0)e^{-t/\tau}$$

Here τ is a characteristic time defined as $\tau = C/k$, where C is the effective heat capacity of the object and k is the constant introduced above in Newton's cooling law. But Newton's cooling law is only an approximation. When the temperature difference $T_c - T_0$ is large, the cooling rate usually varies faster than linearly in $T_c - T_0$. Furthermore, there is no reason to assume that the constant k is the same whether milk is added or not. For instance, the evaporation may be affected, and the geometry of the system is changed, when more liquid is added. It seems obvious that theoretical arguments based directly on Newton's cooling law give uncertain conclusions. Experiments have shown that adding the cold milk immediately reduces the cooling rate somewhat, with the net effect that the coffee is warmer (compared with the case when milk is added later) after, say, 15 min. The main reason seems to be the reduced evaporation when milk is added. If we really want to increase the cooling rate of a much-too-hot coffee, we could stir it frequently and rely on the large cooling effect from evaporation and convection. Whether the milk is added immediately or later is not very important. If the issue is to keep the coffee hot for a while, the best is to put on a lid to prevent convection and evaporation. The time when milk is added is not important.

5.7 Time for Contact

The contact time is much longer than 1 μs.

This is an example where an anthropomorphic approach can be helpful. (*anthropomorphic:* attributing a human personality to anything impersonal or irrational; Oxford English Dictionary.) Imagine that you sit on the falling sphere, opposite to the collision point. Your friend sits on the other sphere, and at the collision point. When "your" sphere collides, you cannot know that it has hit an obstacle until a "signal" has arrived from the collision point to you on the other side. This signal is an elastic wave (a sound wave) through your sphere. If it propagates with the speed u, and the diameter of the sphere is L, it takes the time L/u until you know that the part of the sphere where you sit must start to slow down. It takes another time L/u before a signal has gone from you to the collision point, allowing your friend to know that there is no part of the colliding object further out, which has not yet started to decelerate. In other words, it takes at least the time $2L/u$ before anyone at the collision point can know the size of the incoming object. The spheres must be in contact at least during this time.

This is the first part of the solution. Next we must obtain numbers for L and u. With the information that the mass of the sphere is 1 kg and the density of brass is 8400 kg/m^3, we get the diameter of the sphere $L = 0.061$ m. The propagation speed is less trivial. Perhaps you happen to know the expression

$$c = \sqrt{\frac{E}{\rho}}$$

for the longitudinal sound velocity c in a long bar. The numbers in the table give

$$c = \sqrt{105 \times 10^9 / 8400} \text{ m/s} \approx 3500 \text{ m/s}$$

Taking this value for the speed u gives

$$t_c = \frac{2L}{u} = 35 \cdot 10^{-6} \text{s} = 35 \mu\text{s}$$

This is much longer than 1 μs, which is the time with which we should compare according to the text of the problem. But suppose that you have never heard of the relation for the speed of sound in a long bar. With the data for brass given in the table, you could still make an educated guess. Consider the quantity E/ρ. It has the unit $\text{Pa}/(\text{kg}/\text{m}^3)$. The unit pascal (Pa) is force per area, or N/m^2. Force appears in Newton's equation $F = ma$ as mass times acceleration. When 1 Pa is expressed in only the base units in SI, we get 1 Pa = 1 $\text{kg}/(\text{m}\cdot\text{s}^2)$. The SI unit for E/ρ thus is m^2/s^2, or the square of a speed. Could this be the speed to use in the propagation of a signal in our case? You cannot know for sure, but as argued below it makes a good educated guess. We then conclude that the collision time is not as short as 1 μs. (The collision time can be experimentally determined by recording the time during which a current can flow through the spheres in contact.)

Heinrich Hertz analyzed the problem of two spheres in an elastic collision in 1881. That theory gives a contact time

$$t_H = C\left(1 - \nu^2\right)^{2/5} \left(\frac{c}{w}\right)^{1/5} t_c$$

C is a constant of the order of 1, w is the relative speed of the spheres when they collide, and c and t_c are the quantities calculated above. The expression also contains Poisson's number ν from elasticity theory, a quantity that typically is 1/3.

In our case the sphere was pulled horizontally 0.10 m and then released. The subsequent change from potential energy to kinetic energy gives the speed $w = 0.31$ m/s at impact. Then $(c/w)^{1/5} = 6.5$.

The factor $(1-\nu^2)^{2/5}$ is always close to 1. With these numbers, and $C \sim 1$, we get $t_H \sim 6.5t_c \sim 200$ μs. This is much longer than the time 1 μs we were asked to compare with in the problem.

How physicists think. This problem contains many examples of physics thinking. The anthropomorphic approach in which you imagine yourself being part of the system is not at all childish. Experienced physicists often use it. In his autobiographical notes, Albert Einstein recalls that he, at the age of 16, wondered what he would observe if he could pursue a beam of light with the velocity c of light in a vacuum.

Dimensional analysis refers to considerations based only on the dimensions (units) of those quantities that one assumes are relevant for a certain problem. Our use of this approach, to estimate the speed of sound in brass from the given data, exemplifies an important aspect of how physicists think. It could not give a precise value for the collision time, but only an educated guess. The comparison with the accurate result, obtained through *Hertz's formula,* showed that the guess was quite adequate for our task, to ascertain that the contact time is longer than 1 μs. In problem 4.8 (Rise and Fall of a Ball) we discussed at some length the concepts of a characteristic length and a characteristic velocity. In the present problem, the quantity

$$t_c = 2L\sqrt{\frac{\rho}{E}}$$

plays the role of a *characteristic time.* Another characteristic time is the contact time calculated from Hertz's theory (the symbol ~ means "of the order of")

$$t_H \sim \left(\frac{c}{w}\right)^{1/5} t_c$$

Our discussion showed that $t_H \gg t_c$. Note that whereas dimensional arguments give a unique combination of the powers of L, ρ, and E

that enter t_c, we cannot use such a simple approach in the case of t_H, which involves two more quantities, c and w. Physical modeling, and not only dimensional reasoning, is necessary to obtain t_H.

Any caveat? The contact time discussed here holds for elastic collisions, that is, with no permanent deformations of the colliding bodies. It should be a very good approximation for brass spheres colliding with a low relative speed. It is obvious that Hertz's formula is useless, for example, for the collision of two spheres of wet clay.

Outlook. The familiar Newton's cradle (fig. 5.3) also deals with suspended and colliding spheres. Pull one sphere to the side and release it. One sphere is knocked out at the other end of the row while the rest of the spheres are not moving. With two spheres pulled to the side and released, two spheres are knocked out, and so on. This behavior is sometimes explained as the result of the successive collision of two spheres at a time, that is, one assumes that there is a small separation between each sphere. Although that description may not be wrong in certain cases, it is an unnecessary simplification as is discussed in references listed in "Further Reading."

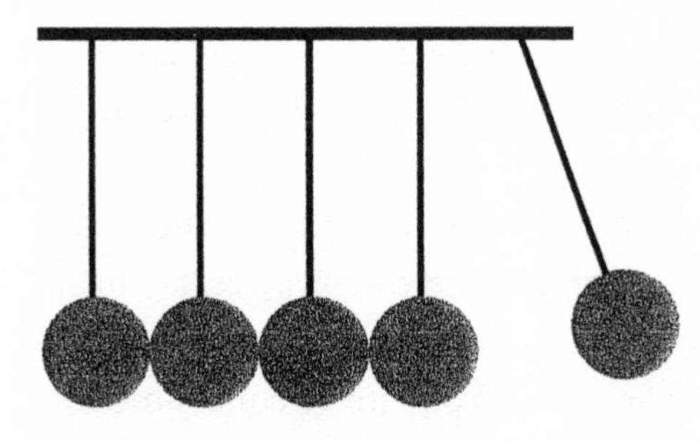

Fig. 5.3. Newton's cradle

5.8 Socrates' Blood

The surprising answer is that one is not "allowed" to ask this question in physics. A numerical answer would conflict with the second law of thermodynamics.

A typical physicist's approach could be to model the problem as the mixing of the water in Socrates' blood (say 4 liters = 0.004 m^3) with all the water in the world (say 1×10^{18} m^3 of ocean water). Thus $4/10^{21}$ of all water molecules in the world come from Socrates. The glass of water may contain 180 mL, that is, 180 g, or 10 mol. Avogadro's constant 6×10^{23} mol^{-1} now reveals that there are $10 \times 6 \times 10^{23} \times (4/10^{21}) = 24\,000$ molecules from Socrates in the glass. This result is incorrect, but why? Perhaps you object that mixing with all the water in the oceans is too crude an assumption? Water can circulate by evaporation and precipitation. The average annual precipitation over the entire Earth is about 1 m. Socrates died in 399 B.C. and the average depth of the oceans is about 4 km. This suggests that there has not been enough time to mix the water from Socrates with all ocean water. Moreover, in some sense it is the same top layer of the oceans that takes part in the hydrologic cycle. If so, our estimate is much too low and a more reasonable answer may be of the order of 10^6 molecules. We should reduce that estimation somewhat because water in the oceans can also circulate vertically due to density differences, when currents take water to colder regions. But the result for the number of water molecules from Socrates in our glass is still incorrect, and for a very fundamental reason. All water molecules are alike, and cannot be "tagged" like classical macroscopic particles. Therefore, to ask how many molecules in the glass come from Socrates' blood has no physical meaning. These are strong words and require an explanation.

If this had been a legitimate physics question, it would be in conflict with what the second law of thermodynamics says about entropy. Consider the well-known textbook case in which a box is divided into two identical compartments, containing equal amounts of nitrogen molecules (fig. 5.4). A hole is opened in the partitioning wall, and molecules are allowed to pass through. If the hole is then closed, the final state of the gas is indistinguishable from the initial state. There has been no mixing in a physical sense, that is, no increase in disorder and hence no increase in entropy. To ask how many molecules from the left compartment are afterward in the

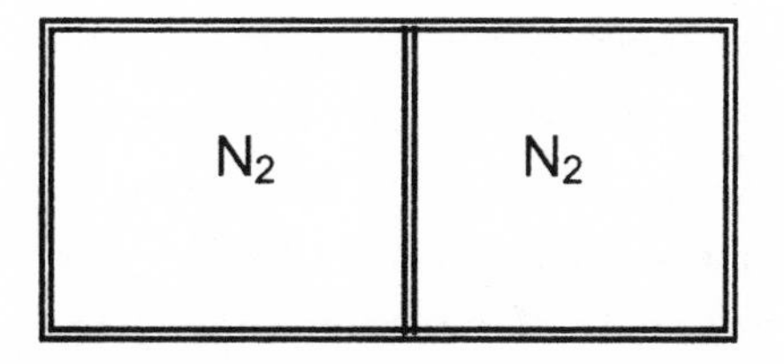

Fig. 5.4. A box with two compartments contains nitrogen molecules. Do these molecules mix if a hole is opened in the wall that separates the compartments?

right compartment has no physical significance. In classical statistical physics we encounter this dilemma when certain expressions referring to N identical particles must be divided by $N!$ (i.e., N factorial).

If you are suspicious about these arguments, the following analogy may be helpful. If you travel from the United States to Europe and buy euros with a 100-dollar bill at a European airport, a certain probability exists that this particular dollar bill will later be in your hands again. One could work out a model and check the number printed on the bill. But if you use your credit card when you buy euros, and if you later get 100 dollars transferred electronically to your bank account, it has no meaning to ask what is the probability that the 100 dollars you paid in Europe are exactly those 100 dollars that were later paid to your account.

How physicists think. Physics deals with a description of Nature that can be experimentally tested, or at least observed as in cosmology. Nature sometimes behaves in such a way that certain questions cannot be answered, even in principle. The particle-wave duality in the double-slit experiment may be the best-known example. We cannot insist on asking which slit the electron has passed through.

The so-called *Fermi problem* (estimation problem) considered here can be found in numerous physics books, in one form or another, and is usually given an incorrect answer in the form of an estimated numerical value. For instance, how long would it take for an electron to pass through the wires in a house with DC electricity. Because of the indistinguishability of the electrons this is not a physically relevant question, despite the fact that one could easily work out an answer in terms of the velocity and the mean free path of the electrons, and the length of the wiring in the house.

Any caveat? Water molecules are not stable entities because they take part in photosynthesis. That might be a complication if we maintain that the problem has a meaning in a statistical sense. But we could consider another problem that has the same character. How many argon atoms (which are stable) do we now inhale that were also in Socrates' last breath? Furthermore, according to Plato, Socrates was given poison hemlock. We might wonder how many molecules of the poison there are in our glass of drinking water. This would be a legitimate question only if the poison had been unique, with no other molecules of that sort existing in the world, earlier or later, besides those given to Socrates. As another example, assume that certain amounts of a new stable compound are released into the atmosphere from several different sources. One may then ask how much of this compound one expects to find later in a sample of the atmosphere, as long as the *individual* molecules found are not identified as coming from a particular source.

ADDITIONAL CHALLENGES

5.5 Grains of Sand

We found that a "sand heap" with the volume of 10^{11} m^3 would have approximately as many grains of sand as there are atoms in a single grain. Such a sand heap corresponds to a 10-m (30-ft)-deep layer of sand covering a square with side lengths 100 km $\times$ 100 km (60 miles $\times$ 60 miles), that is, not an exceptional size for a desert.

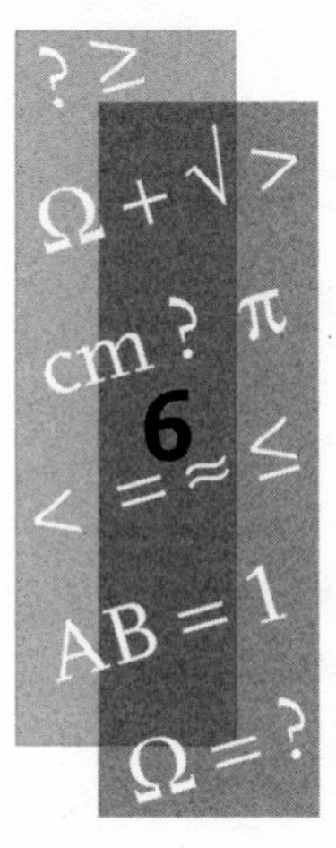

6 Challenges for Your Creativity

This chapter contains problems in which the solution requires an element of creativity before the knowledge of physics is applied.

PROBLEMS

6.1 Iron Bars

Two bars, painted black and with dimensions 1 cm × 2 cm × 10 cm, lie on a table. They look identical but one is made of nonmagnetized iron and the other is made of iron that is permanently magnetized, thus making it a bar magnet. How can you decide which one is a magnet, using nothing but your hands?

6.2 Faulty Balance

With a standard set of accurate weights, and with a perfect balance, one can determine the mass of an object. You have the set of weights, but unfortunately the balance is faulty. Without anything placed on the scales there seems to be equilibrium, but you know that the arms of the balance are not of equal length. This has been compensated for by adjusting the weight of the scales. You can still determine the mass of an object. How is it done?

6.3 Greek Geometry

Euclid was a Greek mathematician (c. 325–c. 270 BC) whose works dominated geometry for more than 2000 years. Many famous mathematical problems cannot be solved by Euclidean geometry. One of these problems is the trisection of a plane angle, another is the "squaring of the circle" (the construction of a square that has the same area as a given circle, which means finding π), and a third is the "doubling of a square" (finding the side length of a square whose area is twice that of a given square, or the square root of two).

In strict Euclidean geometry one is allowed to use only a ruler (without grading, i.e., a straightedge) and a pair of compasses to solve a problem. Our task is even more constrained. You are allowed to use only the ruler—and only to draw straight lines with a pencil. How can you determine the position of the center of mass of a thin L-shaped form by a graphical construction (fig. 6.1)?

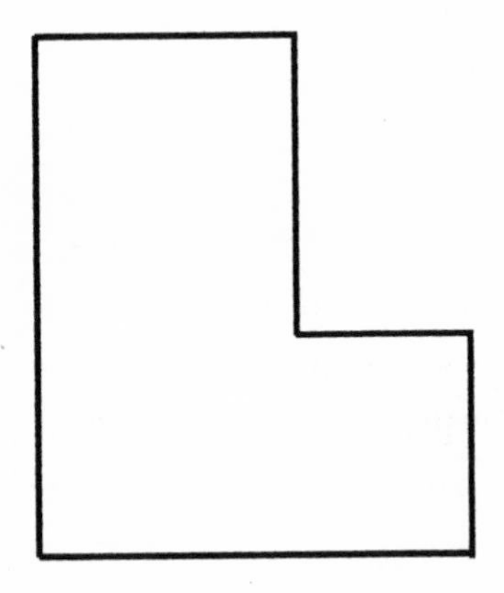

Fig. 6.1. Find the center of mass of this figure by a graphical construction, using only a straightedge and a pencil.

6.4 The Sugar Box

An open, but full, ordinary cardboard box of lump sugar stands on your kitchen table. How can you determine the static friction factor between the table and the box with the help of a pencil? The error should be less than 15 %. You are not allowed to tilt the table.

6.5 The Catenary

In this problem we compare the location of the center of gravity of two hanging objects.

First we take two thin, stiff bars, which form a V-shaped figure. It hangs freely from points A and B, as in figure 6.2. The center of gravity (CG) is located half-way down from the horizontal line connecting A and B to the lowest point (C) of the bars. Then we take a chain whose length is equal to the total length of the two bars in the V-shaped figure. When the chain hangs freely from A and B it has the so-called catenary shape. Is the center of gravity of the chain located above, below, or at the same height as for the V-shaped figure?

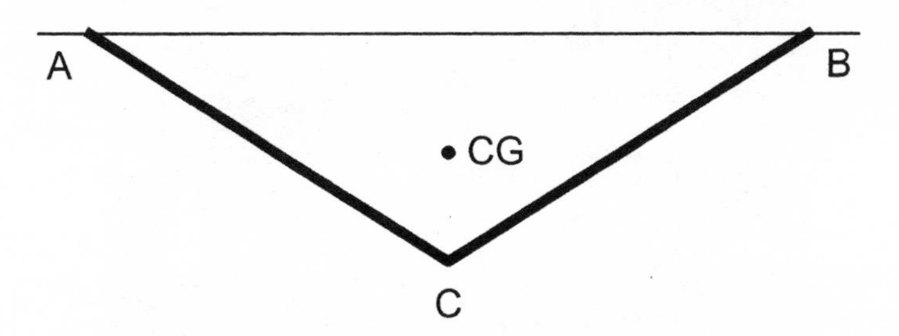

Fig. 6.2. Would the center of gravity (CG) lie higher, lower, or at the same level if the two black bars are replaced with a chain of equal length?

6.6 False Impressions

Take a close look at the picture (fig. 6.3). Can you find something that is not according to the laws of physics?

6.7 Testing the Hammer

The International Association of Athletics Federations (IAAF) gives very detailed rules for the implements used in track and field sports. In hammer throwing, the implement has three parts: a metal head, a wire, and a grip. The head is a sphere, although not quite so because there must be something to attach it to the wire. Rule 191 says about the head: "The centre of gravity shall not be more than 6 mm from the centre of the sphere."

Fig. 6.3. Find at least three features in the picture that are not according to a law of physics.

How can this requirement for the center of mass be easily checked—not in a laboratory but at the competition site?

6.8 Which Way?

Figure 6.4 shows a track left by a bicycle. Did the bicycle move to the right or to the left? (The unequal length of the tracks can be explained because the wheels passed through different patches of wet paint before they left the marks on the road.)

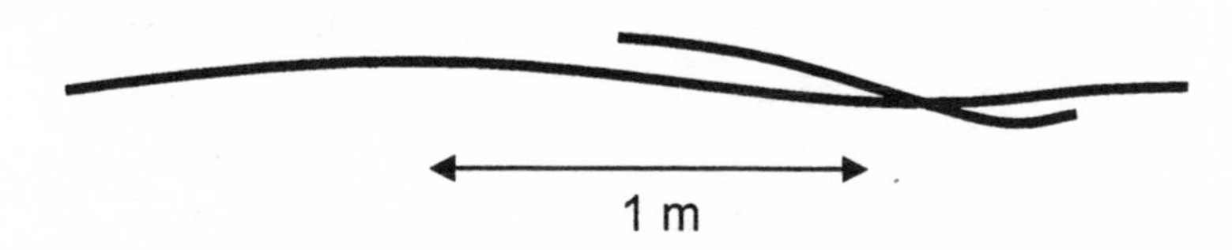

Fig. 6.4. Tracks left by a bicycle. In which direction did the bike go?

6.9 Three Switches

In the first floor of a building there are three switches labeled 1, 2, and 3, and with on/off indicated correctly. In the basement corridor there are three ordinary lightbulbs labeled A, B, and C (fig. 6.5). Each switch in the first floor is connected to one (and only one) of the lightbulbs in the basement. There are no other connections between

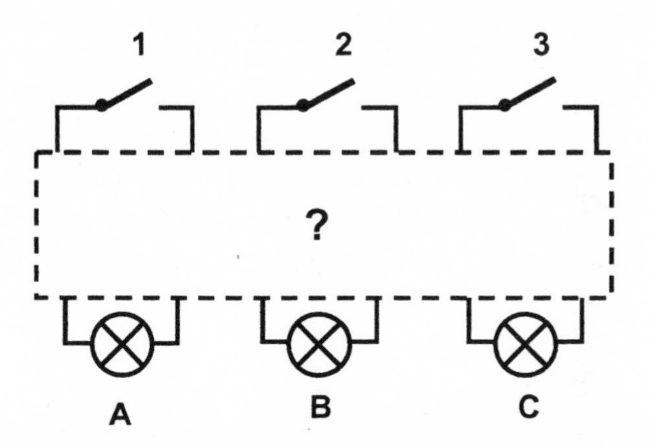

Fig. 6.5. Three switches are connected with three lightbulbs, but how?

the three bulbs or between the three switches. It is your task to decide which switch operates bulb A, B, and C, respectively. You are allowed to go down to the basement from the first floor, but only once. After that you must be able to tell how the wiring is done.

6.10 Pulse Beats

You have been out jogging, a strenuous exercise. Afterward you take a slow swim in the lake, and just relax. Then you want to see if your pulse is back to its normal value, but you have no watch. Knowing a little bit of physics you soon find out a way to solve your measurement problem. What do you do?

6.11 Fake Energy Statistics

The total energy consumption in a country is usually expressed in Mtoe (megatons of oil equivalents). Table 6.1 gives this value for the 25 countries that are currently (2006) members of the EU (European Union). One of the columns refers to actual data (from 2003) and the other column contains fake values. Which is the correct column?

SOLUTIONS

6.1 Iron Bars

In a bar magnet one end is a magnetic north pole and the other end is a magnetic south pole. The middle of the bar is neutral with respect to magnetic poles. Let one bar remain on the table and hold the other bar vertically so that it touches the middle of the bar on the table (fig. 6.6). If there is an attraction between them, it is the nonmagnetized bar that lies on the table.

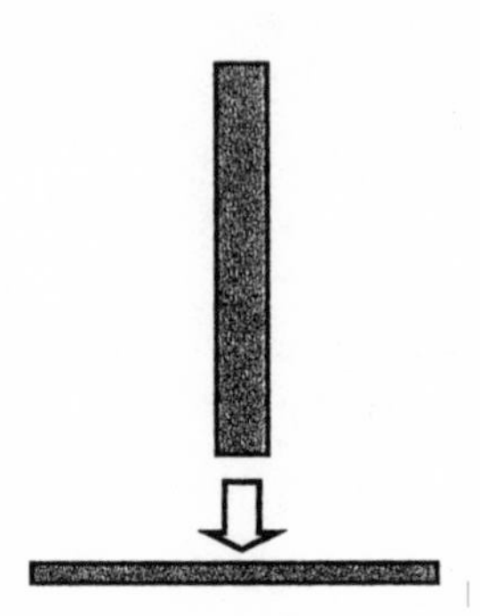

Fig. 6.6. A bar magnet lifts a piece of iron, but a piece of iron does not lift a bar magnet, if placed as in the figure.

Table 6.1. Energy consumption in 2003 by final consumers in the 25 EU countries. The countries are ordered alphabetically according to the official two-letter abbreviation (e.g., BE for Belgium).

Country	Mtoe	Mtoe
Belgium	38	62
Czech Republic	25	55
Denmark	15	22
Germany	231	412
Estonia	3	8
Greece	20	51
Spain	90	98
France	158	220
Ireland	11	22
Italy	130	198
Cyprus	2	3
Latvia	4	8
Lithuania	4	9
Luxembourg	4	4
Hungary	18	50
Malta	0.5	0.9
Netherlands	52	97
Austria	26	61
Poland	56	94
Portugal	18	32
Slovenia	5	8
Slovak Republic	10	20
Finland	26	31
Sweden	34	71
United Kingdom	150	219

Mtoe = megatons of oil equivalents

6.2 Faulty Balance

The mass is given by the square root of the results of two weighings, with the unknown mass first on one side and then on the other side of the balance.

In mathematical terms, the argument is as follows. The unknown mass is m, and the lengths of the arms of the balance are L_1 and L_2. When the unknown mass is placed on the left side, it is balanced by a mass m_2 on the other side. When placed on the right side, it is balanced by the mass m_1. We have

$$mL_1 = m_2L_2$$

$$m_1L_1 = mL_2$$

Putting $L_2 = m_1L_1/m$ from the last equation into the right side of the first equation gives

$$mL_1 = m_2(m_1L_1/m)$$

or

$$m^2 = m_1m_2$$

Even if the balance is obviously faulty and has arms tilting to one side when nothing is placed on the scales, we could use it to determine an unknown mass. Just put weights on one of the scales to restore an apparent equilibrium. We don't need to know the mass of these weights. Grains of sand, or anything else such as some of the pieces in our set of weights, could be used to get the situation where we can apply the mathematical argument above.

Additional challenge. How many pieces of weights, whose masses in grams are given by integer numbers, are needed to measure masses ranging from 1 g to 1 kg, with an error of at most 1 g? The balance itself is assumed to be very accurate. (See the solution at the end of this chapter.)

6.3 Greek Geometry

The L-shaped figure can be divided into two smaller rectangles (fig. 6.7). Each small rectangle has its center of mass (CM) where its diagonals cross. The CM of the two combined rectangles lies on the straight line AB going through the CM of each of the smaller rectangles. But the L-shaped figure can be divided into rectangles in two ways (fig. 6.8). Repeat the construction for the other way. The CM of the L-shaped figure lies where the two straight lines in the two constructions cross.

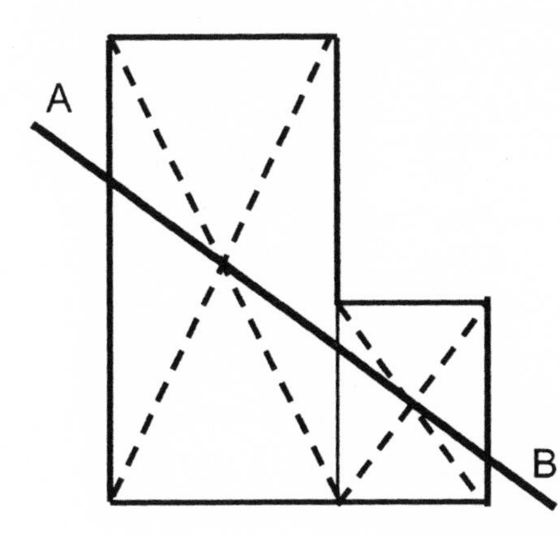

Fig. 6.7. The center of mass of the L-shaped figure lies on the line AB.

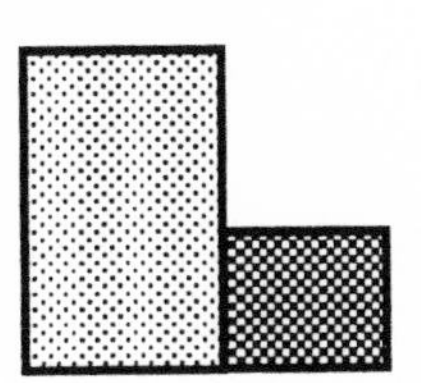

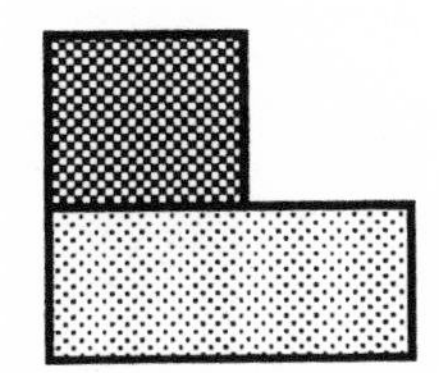

Fig. 6.8. Two ways to divide an L-shaped figure into rectangles

In the construction just suggested, the two lines may cross at a rather small angle. That can make it difficult to find the precise position of the crossing point with high accuracy. It would be much better if the two crossing lines make a large angle. We can achieve this with a different construction. Complete the L-shaped figure so that it forms a rectangle. We now have one large and one small rectangle, whose CMs are at their centers. The CM of the L-shaped figure lies on the line through the CM of the two rectangles (fig. 6.9).

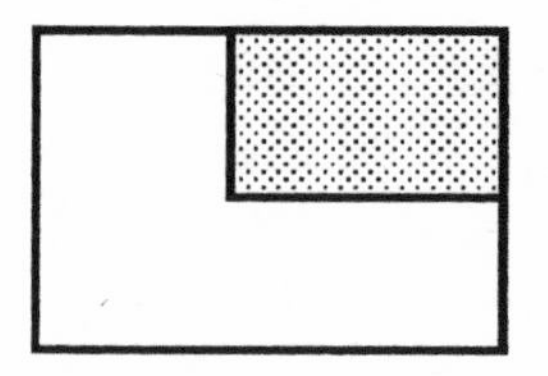

Fig. 6.9. An L-shaped figure plus a rectangle form a new rectangle.

6.4 The Sugar Box

Push horizontally on the side of the box, with increasing force, until it either starts to slide or to tilt. Which one of these alternatives that happens first depends on how high up on the box one pushes. Try pushing at increasing height h until the box just starts to tilt. With the geometry in figure 6.10, the (unknown) force F then gives the torque Fh counterclockwise around point O. It is balanced by the torque in the opposite direction from the weight Mg. We have

$$Fh = Mga$$

The normal force is $N = Mg$ and hence the fully developed friction force is $fN = fMg$. The friction force is balanced by the horizontal force F:

$$F = fN = fMg$$

Combining these relations gives the friction factor

$$f = \frac{F}{Mg} = \frac{a}{h}$$

It remains to determine the ratio of two lengths, a/h. Because the ratio is independent of the length unit used, we may, for instance, measure a and h in units of the size of a piece of lump sugar.

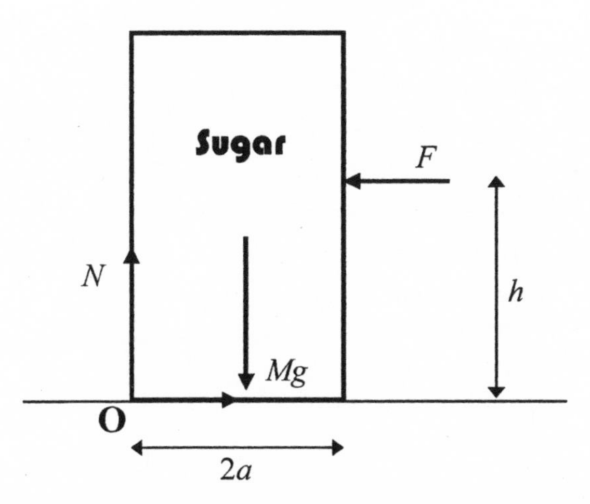

Fig. 6.10. Forces acting on the box when it is just about to tilt counterclockwise

6.5 The Catenary

The chain has the lower center of gravity.

If you pull the middle of the chain strongly downward, it forms exactly the same shape as the V. The stretched chain and the V then have centers of gravity at the same level. Now release the pull on the chain. It spontaneously returns from the V shape to the catenary shape. This means that the potential energy decreases, that is, the center of gravity of the chain sinks to a lower level.

The general shape of a catenary is the same as that of the function

$$\cosh(x) = \frac{e^{x} + e^{-x}}{2}$$

This is the form that the cables in a suspension bridge would take if there were no load from the bridge itself. With a load that is evenly distributed along the horizontal direction, the cables form a parabola.

6.6 False Impressions

Some of the reflections in the water are not correct, because they show objects that should be hidden behind other objects. For instance, consider part of the chimney, or the root of the tree near the water. Did you find more dubious reflections?

The schematic illustration in figure 6.11 shows how an observer may see an object directly but not reflected in the water.

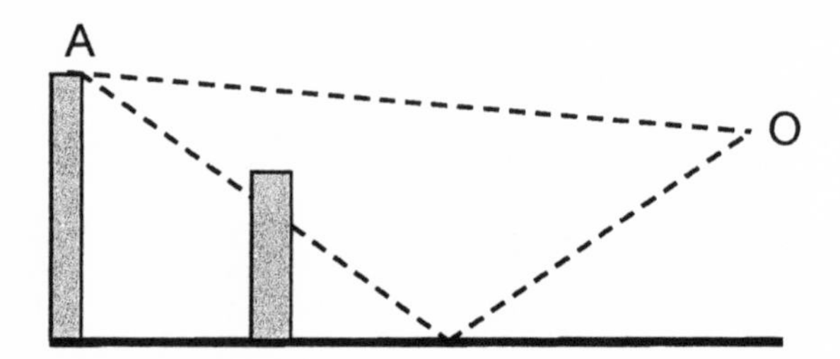

Fig. 6.11. Observer O can see A directly, but not as a reflection.

Additional challenge. One of the most famous paintings by Diego Velásquez is *The Rokeby Venus,* which hangs in the National Gallery in London. You may easily find this picture on the web. Take a look at it. What does the woman see in the mirror? (See the solution at the end of this chapter.)

6.7 Testing the Hammer

The IAAF rules give an ingenious solution: "It must be possible to balance the head, less handle and wire, on a horizontal sharp-edged circular orifice 12 mm in diameter" (fig. 6.12).

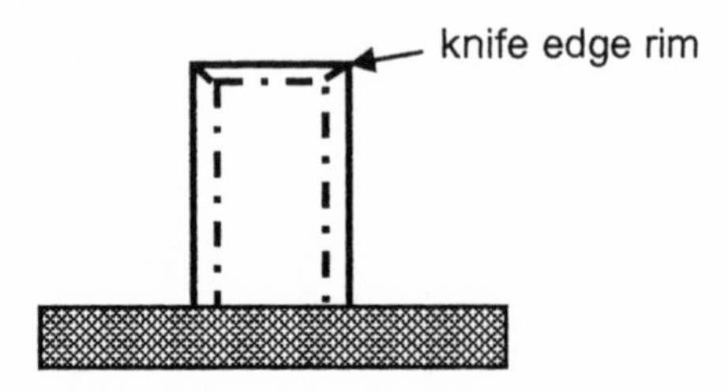

Fig. 6.12. The hammer head should balance on a circular orifice, 12 mm in diameter.

Because the head is spherical, but at the same time it is not a perfect homogeneous sphere because of the connection device between the head and the wire, one should orient the head in many ways on the sharp-edged orifice. If it never falls down, the center of gravity lies within 6 mm of the geometrical center of the spherical head, as required by the rules.

As an example of how detailed the IAAF rules are, we quote the following from rule 191. "The head shall be of solid iron or other metal not softer than brass or a shell of such metal, filled with lead or other solid material. If a filling is used, this shall be inserted in such manner that it is immovable and that the centre of gravity shall not be more than 6 mm from the centre of the sphere. The wire shall be connected to the head by means of a swivel, which may be either plain or ball bearing."

6.8 Which Way?

The front wheel makes wider and less smooth turns than the rear wheel. It is likely, therefore, that the short track is from the front wheel. Next, we note that the rear wheel is always lined up with the frame of the bicycle, but the front wheel often makes an angle relative to the frame. Furthermore there must be a constant distance between the two points where the front and rear wheels make contact with the road. Draw a tangent to the track of the rear wheel, and in the assumed direction of the motion of the bike. This tangent always coincides with the orientation of the bicycle frame. The distance along the tangent to the point where it crosses the track of the front wheel must be equal to the distance between the wheels, typically about 1 m. Try some tangents in the figure. It is obvious that only a bike moving to the right in the figure can give rise to such tracks.

6.9 Three Switches

Let switch 1 first be on and then, after a while, off again. Switch 2 is switched on permanently, and switch 3 is left in the off position.

When you come to the basement you will see one bright lightbulb. It is connected to switch 2. When you touch the two other bulbs you find that one of them is warm. That bulb is connected to switch 1.

Additional challenge. When you come to the basement you find that no lightbulb is lit, but one of them is warm. Can you still figure out the wiring, without returning to the first floor? (See the solution at the end of this chapter.)

6.10 Pulse Beats

All jogging shoes have shoestrings. You can make a pendulum with your shoe (or some other object) hanging on the shoestring. Its frequency depends on the length of the string, as described mathematically below. The length itself can be easily determined if you use the fact that the distance between the fingertips of your outstretched hands is very close to your own height.

We start with a well-known result for an ideal pendulum (also called a mathematical pendulum). All its mass m is concentrated in one point. If its length is L, the period, that is, the time for a complete oscillation back and forth, is

$$\tau = 2\pi\sqrt{\frac{L}{g}}$$

From elementary physics courses it is well known that τ depends only on L and on the acceleration of gravity $g = 9.8\ \mathrm{m/s^2}$, but not on the mass m or on the amplitude of the pendulum, as long as the amplitude is small. An ideal pendulum with $L = 1.00$ m has $\tau = 2.01$ s when $g = 9.8\ \mathrm{m/s^2}$. It is not difficult to obtain this result by putting numbers into the formula, but if you are a physicist you may have found it useful to remember that "a mathematical pendulum with the period 2 s has the length 1 m." Of course this is only an approximate result, which rests on the fact that $\pi^2 \approx 9.87$ is close to the numerical value of the acceleration of gravity, $g = 9.8\ \mathrm{m/s^2}$.

How physicists think. How accurate is a determination of the pulse rate, as described above? No experimental value should be given without an estimation of its accuracy. There are at least three aspects to consider in our case. Perhaps you may have tied the upper end of the shoestring to a tree branch and measured your pulse while it swings. Is it possible to consider the amplitude as small? Second, a jogging shoe hanging on a shoestring is not exactly a mathematical pendulum. Does it matter? Finally, the pendulum motion is damped because of energy losses. How wise is it to take a formula that ignores such effects?

To answer the first question we seek a correction in terms of the maximum angular displacement θ from the vertical. We expect a series expansion such that the correction to the period of the ideal pendulum is expressed in a power of the small angle θ. In advanced textbooks on classical mechanics one may find that (with θ in radians)

$$\tau \approx 2\pi\sqrt{\frac{L}{g}\left(1+\frac{1}{16}\theta^2\right)}$$

If we let θ at the start be less than about 0.17 radians, corresponding to the angle 10°, we can safely neglect the correction due to the finite amplitude. It changes τ by less than 0.1 %.

Next we consider the fact that a jogging shoe does not have all its mass concentrated in a single point, as required for a mathematical pendulum. Upper and lower limits to this effect are obtained if we let L be the distance to either the highest or the lowest parts of the shoe, when it hangs vertically on the shoestring. Typically, these limits correspond to a variation in L by, say, ±4 %, which gives ±2 % variation in τ.

The effect of energy loss (damping), leading to a gradual decrease in the angular displacement, is more complicated to analyze. It arises, for example, from the air resistance acting on the shoe as it swings. A more detailed analysis of a damped pendulum (not given here; see "Further Reading") suggests that its direct effect is not important in our case. An indirect correction arises because the maximum angular displacement gradually decreases, but we have just noted that the amplitude is not of much concern in our case. Furthermore, if the damping of the swinging shoe is too large, you may want to give the shoe an extra push a few times. If this push is synchronized with the swing, the correction to the measured pulse rate is small compared with several other sources of error, although it is difficult to estimate.

Putting it all together, we find that the makeshift clock is remarkably accurate, and should allow you to determine your pulse rate to better than ± 2 beats per minute. That should be sufficient for your needs.

6.11 Fake Energy Statistics

The column to the right is a fake. The numbers in the two columns are rather close, so it may not be easy to decide which column is a fake just by trying to make an estimation for each country. But if we look at how the nine digits 1, 2, . . . , 9 are distributed among the first digits in each column, we find two quite different patterns. Table 6.2 shows the actual number of entries beginning with 1, 2, etc. in the left (true) and in the right (fake) column, and also the expected number according to Benford's law (see below). For instance, there should be many more entries beginning with digits 1 to 4 than with digits 6 to 9. Because we have only 25 countries on the table, there will be deviations from the law.

An argument for Benford's law is as follows. Note that entries beginning with 9 or 8 can differ by at most a factor of 9.999 . . . /8.000 . . . = 1.25, while entries beginning with 2 or 1 can differ by at most a factor of 2.999 . . . /1.000 . . . = 3. There is much more room for data in the latter interval. As another argument, suppose that we change the unit (in our case Mtoe) to another energy unit (e.g., TWh). All entries will be multiplied by the same factor. Only a logarithmic distribution for the occurrence of the digits 1 to 9 as the first digit is invariant under such a change of the unit. Of course the

Table 6.2. Actual number of entries in the left and right columns of table 6.1 and the number expected according to Benford's law. Digits 1–4 should be more common than 6–9 as the first digit.

Value of first digit	1	2	3	4	5	6	7	8	9
Occurrence in left column (true)	8	6	3	3	4	0	0	0	1
Occurrence in right column (fake)	1	5	3	2	3	2	1	3	5
Expected occurrence, Benford's law	8	5	3	2	2	2	1	1	1

law only holds if we consider a large volume of data, which covers several orders of magnitude. It does not hold, for example, for the length of human lives (be they expressed in years, or months, or days), because normally that length does not vary much. You can test Benford's law on the web by using a search engine to look for freely chosen numbers, e.g., 130, 560, and 910, and see how many hits you get for these numbers. There should be many more for 130 than for 910.

Credit for the rule is often given to the American physicist Frank Benford who, in 1938, published an extensive study. He analyzed numbers on utility bills, lists of prices for various goods, areas of geographical regions, street numbers in the addresses of members of the American Physical Society, and many other collections of data—altogether 20 229 numbers. Almost one third of all numbers have 1 as the first digit. The rule is sometimes called *Benford's law,* or the *significant-digit law.* An interesting application is in the detection of forged data in economic transactions. For instance, if one is freely inventing a large number of expenses, and is not aware of Benford's law, it is natural to let the first digit in the cost of each item be a number chosen randomly between 1 and 9.

The uneven distribution of the value of the first digit was not first discovered by Benford, but had been known at least since the second half of the nineteenth century. Simon Newcomb (1835–1909) was a professor of mathematics and astronomy at the Johns Hopkins University, with a particular interest in astronomical data. He published his results in the *American Journal of Mathematics,* the oldest mathematics journal in the Western Hemisphere in continuous publication, which the Johns Hopkins University Press was founded to issue. In a paper from 1881 Newcomb noted that in books of logarithmic tables, the pages for the logarithms of numbers beginning with 1 were more worn than pages for numbers beginning with 9. Tables of logarithms were used to facilitate the calculation of products of numbers, in particular in astronomy. Obviously this was long before even mechanical calculators had been invented. Simon Newcomb was not only a great mathematician, having received many awards, but he also published many popular books on astronomy and even a science fiction novel, *His wisdom the defender* (1900).

ADDITIONAL CHALLENGES

6.2 Faulty Balance

With ten weights, having the values (in grams) 1, 2, 4, 8, 16, 32, 64, 128, 256, 512, we can form combinations that cover all integer values of masses from 1 g to 1023 g. With the same set except for the weight of 1 g, we can form all even-numbered masses 2, 4, 6, . . . , 1022 g. If the measurement of an actual mass is always reported as an odd number of grams, we are at most wrong by 1 g. For instance, if the mass is $m = 213.7$ g, the weighing gives $212 \text{ g} < m < 214 \text{ g}$. We then report the result as 213 g. The required minimum set of weights is obviously related to the digital system of numbers. It contains fewer weights than in a standard set which may have the values (in grams) 1, 2, 2, 5, 10, 20, 20, 50, 100, 200, 200, 500, i.e., *twelve* weights instead of ten, to cover all integer values from 1 to 1000 (actually, to 1110). No doubt a standard set is more convenient in practice, and it requires only two additional weights. But we can also manage with fewer than ten weights. Consider the set 1, 2, 7, 21, 63,

189, 567, 1701. Each number in the sequence equals two times the sum of the preceding numbers plus 1. Now we can form all integer values from 1 to 2551 with eight weights. For instance, the mass 18 is obtained by putting 21 on one scale and 1+2 on the other. To get 500 we put 567 + 1 + 2 = 570 on one scale and 63 + 7 = 70 on the other. If we are content with only even numbers, as in the binary example above, we can take the sequence 2, 6, 18, 54, 162, 486, 1458. (Can you see how this sequence is formed?) It allows us to weigh all masses from 0 g to 2186 g with an error less than 1 g, and using not more than seven weights.

6.6 False Impressions

She sees the painter (or the viewer), but not her own face. There are many famous paintings where the laws of optics are not strictly obeyed. One example is *A bar at the Folies-Bergère* by Édouard Manet, which hangs in the Courtauld Gallery in Somerset House, London. But physicists should be aware that artwork is not always intended to give a correct visual representation. Just think of paintings by Pablo Picasso.

6.9 Three Switches

At least one of the lightbulbs must be broken, but it could not be the warm bulb. You know that the latter is connected to switch 1. Unscrew the warm bulb and let it replace one of the cold bulbs. If it lights up, you know that it is connected to switch 2. Otherwise it is connected to switch 3.

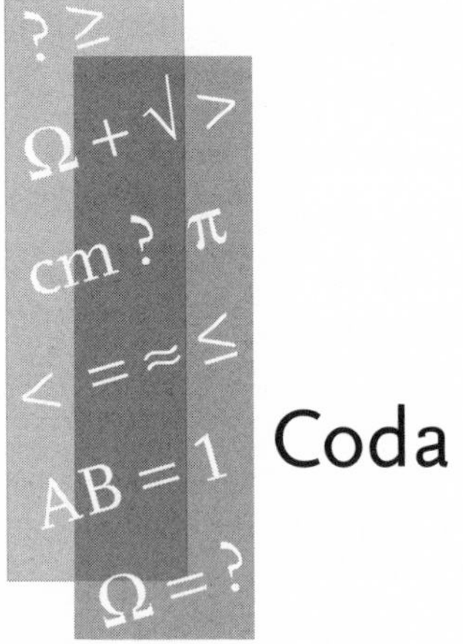

Coda

A main theme in this book has been to use challenging problems to illustrate how physicists think. In schoolbooks the end-of-the-chapter problems are almost always to be solved through a direct application of the theory and principles just dealt with. This is an important training that should not be frowned upon. But real life may not be as simple as those problems assume. What kind of training is then needed to avoid making mistakes? One of the trickiest sources of errors is the presence of an instability. An argument firmly based on physical principles is valid only if those principles can be applied to the situation at hand. It may be difficult to realize that this is not the case. But having encountered many examples of instabilities can lead to a humble attitude and a mental readiness to accept that one may have reached a wrong conclusion.

This book contains several examples of unexpected results. It also contains many examples of very useful and more straightforward physics thinking. Such thinking can be summarized in the following four points, which are in the arsenal of experienced physicists when they attack a problem.

1. Is the result dimensionally correct? Everyone who performs mathematical calculations makes mistakes, now and then. The most important technique to detect such errors is to

check that all expressions have the correct physical dimension.

2. Is the result correct, or at least reasonable, in various limits such as very high (even infinite) or very low (even zero) speeds, masses, densities, distances, and so on? This is the second important technique to detect errors in mathematical expressions.
3. Does it really matter? In other words, could it be that an effect is so small that it has no practical importance, or that it is swamped by other effects? Then one should not waste time and effort on it.
4. If one ends with a numerical result—is it reasonable? A comparison with several other ways of estimating the desired quantity may lead to the detection of an embarrassing mistake, for instance, a wrong power of ten.

But one should never forget that physics is an experimental science. A theory may be wrong if it does not stand up to the practical test. Provided, of course, that the experiment is not ill conceived or carried out wrongly. Mistakes are always possible, both in theoretical arguments and in experimental tests. One antidote is to practice, and practice again. This book may help to take a small step along that road.

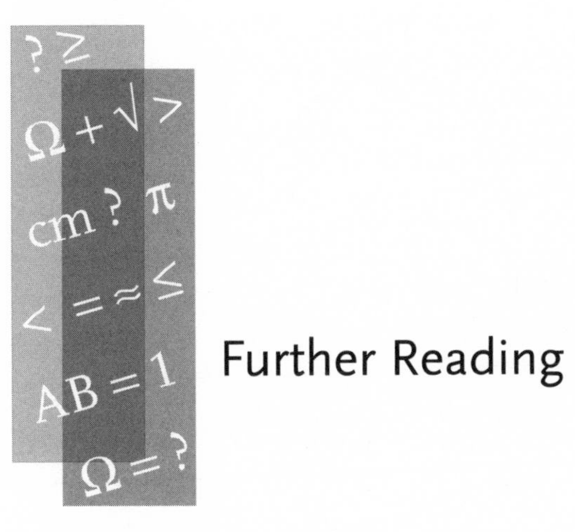

Further Reading

Several of the problems in this book have been treated in scholarly journals, which are mainly devoted to the pedagogical aspects of physics. Works published in *American Journal of Physics* and *European Journal of Physics* are usually intended for teaching at the university level, whereas *The Physics Teacher* and *Physics Education* are concerned with more elementary physics. The works listed here either deal explicitly with the problems we solve or consider aspects that are treated in comments and outlooks.

Abbreviations

Am. J. Phys.	*American Journal of Physics*
Eur. J. Phys.	*European Journal of Physics*
Phys. Educ.	*Physics Education*
Phys. Teach.	*The Physics Teacher*

1.5 Running in the Rain

S. A. Stern. 1983. An optimal speed for traversing a constant rain. *Am. J. Phys.* 51:815–818.

Alessandro De Angelis. 1987. Is it really worth running in the rain? *Eur. J. Phys.* 8: 201–202.

Howard E. Evans II. 1991. Raindrops keep falling on my head. *Phys. Teach.* 29:120–121.

Eileen Scanlon, Tim O'Shea, Randall Smith, Josie Taylor, and Claire O'Malley. 1993. Running in the rain: using a shared simulation to solve open-ended physics problems. *Phys. Educ.* 28:107–113.

J. J. Holden, S. E. Belcher, A. Horvath, and I. Pytharoulis. 1995. Raindrops keep falling on my head. *Weather* 50:367–370.

Thomas C. Peterson and Trevor W. R. Wallis. 1997. Running in the rain. *Weather* 52:93.

1.6 Reaching Out

Paul B. Johnson. 1955. Leaning tower of lire. *Am. J. Phys.* 23:240.

Richard M. Sutton. 1955. A problem of balancing. *Am. J. Phys.* 23:547.

Leonard Eisner. 1959. Leaning tower of the *Physical Reviews*. *Am. J. Phys.* 27:121–122.

R. P. Boas Jr. 1973. Cantilevered books. *Am. J. Phys.* 41:715.

Paul Chagnon. 1993. Animated displays. III. Mechanical puzzles. *Phys. Teach.* 31:32–37.

1.7 Resistor Cube

R. E. Aitchison. 1964. Resistance between adjacent points of Liebman mesh. *Am. J. Phys.* 32:566.

Francis J. Bartis. 1967. Let's analyze the resistance lattice. *Am. J. Phys.* 35:354–355.

Leo Lavatelli. 1972. The resistive net and finite-difference equations. *Am. J. Phys.* 40:1246–1257.

Giulio Venezian. 1994. On the resistance between two points on a grid. *Am. J. Phys.* 62:1000–1004.

F. J. van Steenwijk. 1998. Equivalent resistors of polyhedral resistive structures. *Am. J. Phys.* 66:90–91.

D. Atkinson and F. J. van Steenwijk. 1999. Infinite resistive lattices. *Am. J. Phys.* 67:486–492.

Peter M. Osterberg and Aziz S. Inan. 2004. Impedance between adjacent nodes of infinite uniform D-dimensional resistive lattices. *Am. J. Phys.* 72: 972–973.

1.8 One, Two, Three, Infinity

Robert H. March. 1993. Polygons of resistors and convergent series. *Am. J. Phys.* 61:900–901.

Antoni Amengual. 2000. The intriguing properties of the equivalent resistances of n equal resistors combined in series and in parallel. *Am. J. Phys.* 68:175–179.

S. J. van Enk. 2000. Paradoxical behavior of an infinite ladder network of inductors and capacitors. *Am. J. Phys.* 68:854–856.

R. M. Dimeo. 2000. Fourier transform solution to the semi-infinite resistor ladder. *Am. J. Phys.* 68:669–670.

1.9 Lost Energy

Charles Zucker. 1955. Condensor problem. *Am. J. Phys.* 23:469.

R. A. Powell. 1979. Two-capacitor problem: a more realistic view. *Am. J. Phys.* 47:460–462.

Samuel D. Harper. 1988. The energy dissipated in a switch. *Am. J. Phys.* 56:886–889.

Robert J. Sciamanda. 1996. Mandated energy dissipation—e pluribus unum. *Am. J. Phys.* 64:1291–1295.

William J. O'Connor. 1997. The famous 'lost' energy when two capacitors are joined: a new law? *Phys. Educ.* 32:88–91.

Richard Bridges. 1997. Joining capacitors. *Phys. Educ.* 32:217.

Steven Mould. 1998. The energy lost between two capacitors: an analogy. *Phys. Educ.* 33:323–326.

K. Mita and M. Boufaida. 1999. Ideal capacitor circuits and energy conservation. *Am. J. Phys.* 67:737–739 [Erratum: 2000. *Am. J. Phys.* 68:578].

Sami M. Al-Jaber and Subhi K. Salih. 2000. Energy consideration in the two-capacitor problem. *Eur. J. Phys.* 21:341–345.

A. Gangopadhyaya and J. V. Mallow. 2000. Comment on "Ideal capacitor circuits and energy conservation" by K. Mita and M. Boufaida. *Am. J. Phys.* 68:670–672.

Timothy B. Boykin, Dennis Hite, and Nagendra Singh. 2002. The two-capacitor problem with radiation. *Am. J. Phys.* 70:415–420.

T. C. Choy. 2004. Capacitors can radiate: further results for the two-capacitor problem. *Am. J. Phys.* 72:662–670.

1.10 Simple Timetable

L. K. Edwards. August 1965. High-speed tube transportation. *Scientific American.* 213:30–40.

Martin Gardner. September 1965. *Scientific American.* 213:10–11.

Paul W. Cooper. 1966. Through the Earth in forty minutes. *Am. J. Phys.* 34:68–70.

Philip G. Kirmser. 1966. An example of the need for adequate references. *Am. J. Phys.* 34:701.

Giulio Venezian. 1966. Terrestrial brachistochrone. *Am. J. Phys.* 34:701.

Russell L. Mallett. 1966. Comments on "Through the Earth in forty minutes." *Am. J. Phys.* 34:702.

L. Jackson Laslett. 1966. Trajectory for minimum transit time through the Earth. *Am. J. Phys.* 34:702–703.

Paul W. Cooper. 1966. Further commentary on "Through the Earth in forty minutes." *Am. J. Phys.* 34:703–704.

2.1 Moving Backward?

J. D. Nightingale. 1993. Which way will the bike move? *Phys. Teach.* 31:244–245.

2.4 Low Pressure

Ian Bruce. 1990. Car tyre kinematics. *Phys. Educ.* 25:242.

2.9 Mariotte's Bottle

E. C. Watson. 1939. Edme Mariotte (c. 1620–1684). *Am. J. Phys.* 7:230–232.

J. A. Maroto, J de Dios, and F. J. de las Nieves. 2002. Use of a Mariotte bottle for the experimental study of the transition from laminar to turbulent flow. *Am. J. Phys.* 70:698–701.

3.5 Two Wooden Blocks

Walter P. Reid. 1963. Floating of a long square bar. *Am. J. Phys.* 31:565–568.

R. Delbourgo. 1987. The floating plank. *Am. J. Phys.* 55:799–802.

Paul Erdös, Gérard Schibler, and Roy C. Herndon. 1992. Floating equilibrium of symmetrical objects and the breaking of symmetry. Part 1: Prisms. *Am. J. Phys.* 60:335–345.

Paul Erdös, Gérard Schibler, and Roy C. Herndon. 1992. Floating equilibrium of symmetrical objects and the breaking of symmetry. Part 2: The cube, the octahedron, and the tetrahedron. *Am. J. Phys.* 60:345–356.

Brian R. Duffy. 1993. A bifurcation problem in hydrostatics. *Am. J. Phys.* 61:264–269.

Chris Rorres. 2004. Completing book II of Archimedes's on floating bodies. *The Mathematical Intelligencer* 26:32–42.

3.6 Shot in a Pot

R. C. Johnson. 1997. Floating shells: the breaking and restoration of symmetry. *Am. J. Phys.* 65:296–300.

3.7 Filling a Barrel

Josué Njock Libii. 2003. Mechanics of the slow draining of a large tank under gravity. *Am. J. Phys.* 71:1204–1207.

Richard Humbert. 2005. Water nozzles. *Phys. Teach.* 43:604–607.

3.8 Tube with Sand

Albert A. Bartlett. 1997. The hydrostatic paradox revisited. *Phys. Teach.* 35:288–289.

Haym Kruglak. 1997. Revisiting Pascal's burst barrel. *Phys. Teach.* 35:388–389.

P. G. de Gennes. 1999. Granular matter: a tentative view. *Reviews of Modern Physics* 71:S374–S382.

3.9 Sauna Energy

David J. Smith. 2000. Flexural stress in windows during hurricanes. *Phys. Teach.* 38:400–402.

4.3 The Egg of Columbus

M. E. Gardner. 1966. Falling cylinders. *Am. J. Phys.* 34:822.

4.4 Helium or Hydrogen in the Balloon?

E. C. Watson. 1946. Reproduction of prints, drawings and paintings of interest in the history of physics. 28. The first hydrogen balloon. *Am. J. Phys.* 14:439–444.

4.5 Lightbulb Found in a Drugstore?

Vittorio Zanetti. 1985. Temperature of incandescent lamps. *Am. J. Phys.* 53:546–548.

Dan MacIsaac, Gary Kanner, and Graydon Anderson. 1999. Basic physics of the incandescent lamp (lightbulb). *Phys. Teach.* 37:520–525.

Bruce Denardo. 2002. Temperature of a lightbulb filament. *Phys. Teach.* 40:101–105.

4.8 Rise and Fall of a Ball

Steven Herbert and Terrence Toepker. 1999. Terminal velocity. *Phys. Teach.* 37:96–97.

Paul Gluck. 2003. Air resistance on falling balls and balloons. *Phys. Teach.* 41:178–180.

S. R. Goodwill, S. B. Chin, and S. J. Haake. 2004. Aerodynamics of spinning and non-spinning tennis balls. *Journal of Wind Engineering and Industrial Aerodynamics* 92:935–958.

Jan Benacka and Igor Stubna. 2005. Accuracy in computing acceleration of free fall in the air. *Phys. Teach.* 43:432–433.

5.6 Cooling Coffee

W. G. Rees and C. Viney. 1988. On cooling tea and coffee. *Am. J. Phys.* 56:434–437.

Colm T. O'Sullivan. 1990. Newton's law of cooling—a critical assessment. *Am. J. Phys.* 58:956–960.

Craig F. Bohren. 1991. Comment on "Newton's law of cooling—a critical assessment" by Colm T. O'Sullivan. *Am. J. Phys.* 59:1044–1046.

Michael A. Karls and James E. Scherschel. 2003. Modeling heat flow in a thermos. *Am. J. Phys.* 71:678–683.

5.7 Time for Contact

F. Herrmann and P. Schmälzle. 1981. Simple explanation of a well-known collision experiment. *Am. J. Phys.* 49:761–764.

F. Herrmann and M. Seitz. 1982. How does the ball-chain work? *Am. J. Phys.* 50:977–981.

Bernard Leroy. 1985. Collision between two balls accompanied by deformation: A qualitative approach to Hertz's theory. *Am. J. Phys.* 53:346–349.

Jean C. Piquette and Mu-Shiang Wu. 1984. Comments on "Simple explanation of a well-known collision experiment." *Am. J. Phys.* 52:83.

D. R. Lovett, K. M. Moulding, and S. Anketell-Jones. 1988. Collisions between elastic bodies: Newton's cradle. *Eur. J. Phys.* 9:323–328.

P. Patrício. 2004. The Hertz contact in chain elastic collisions. *Am. J. Phys.* 72:1488–1492.

5.8 Socrates' Blood

G. Grimvall. 2004. Socrates, Fermi, and the second law of thermodynamics. *Am. J. Phys.* 72:1145.

6.10 Pulse Beats

W. P. Ganley. 1985. Simple pendulum approximation. *Am. J. Phys.* 53:73–76.

Robert A. Nelson and M. G. Olsson. 1986. The pendulum—rich physics from a simple system. *Am. J. Phys.* 54:112–121.

Richard B. Kidd and Stuart L. Fogg. 2002. A simple formula for the large-angle pendulum period. *Phys. Teach.* 40:81–83.

L. Edward Millet. 2003. The large-angle pendulum period. *Phys. Teach.* 41:162–163.

Rajesh R. Parwani. 2004. An approximate expression for the large angle period of simple pendulum. *Eur. J. Phys.* 25:37–39.

M. E. Bacon and Do Dai Nguyen. 2005. Real-world damping of a physical pendulum. *Eur. J. Phys.* 26:651–655.

Gerald E. Hite. 2005. Approximations for the period of a simple pendulum. *Phys. Teach.* 43:290–292.

6.11 Fake Energy Statistics

Simon Newcomb. 1881. Note on the frequency of use of the different digits in natural numbers. *American Journal of Mathematics* 4:39–40.

Frank Benford. 1938. The law of anomalous numbers. *Proceedings of the American Physical Society* 78:551–572.

Frank Benford. 1943. The probable accuracy of the general physical constants. *Physical Review* 63:212.

Don S. Lemons. 1986. On the numbers of things and the distribution of first digits. *Am. J. Phys.* 54:816–817.

John Burke and Eric Kincanon. 1991. Benford's law and physical constants: the distribution of initial digits. *Am. J. Phys.* 59:952.

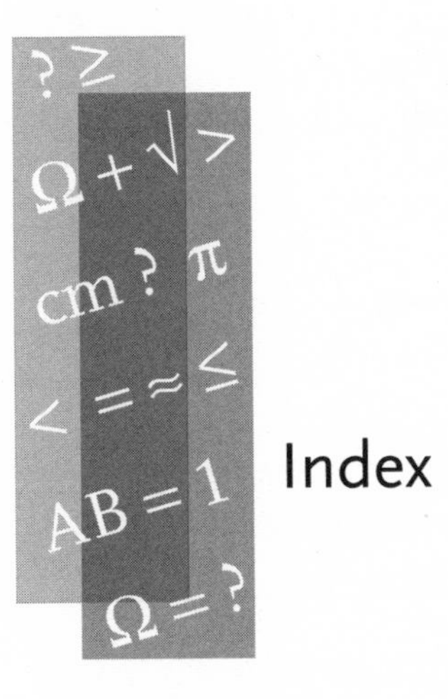

Index